The Earth's Shape and Gravity

A Worden gravimeter in use in the Himalayas, birthplace of the theory of isostasy

The Earth's Shape and Gravity

by

G. D. GARLAND

PERGAMON PRESS

NEW YORK • TORONTO • OXFORD
SYDNEY • BRAUNSCHWEIG

Pergamon Press Inc.
Maxwell House, Fairview Park, Elmsford, N.Y. 10523
Pergamon of Canada Ltd.
207 Queen's Quay West, Toronto 117, Ontario
Pergamon Press Ltd.
Headington Hill Hall, Oxford
Pergamon Press (Aust.) Pty. Ltd.
Rushcutters Bay, Sydney, N.S.W.
Vieweg & Sohn GmbH
Burgplatz 1, Braunschweig

1836/65

Contents

Preface

MEASUREMENTS of the acceleration due to gravity form one of the geophysical methods for investigating the interior of the earth, and also one of the tools of geodesy for the determination of the earth's shape. Several recent developments have led to a great increase of activity in the field: the improvement of instruments, the use of satellites to indicate variations in gravity over the earth, and the availability of high-speed computers to assist in the interpretation of the measurements, to name just three. There has been a demand for a book which would provide at least an introduction to the modern work.

The present volume forms one of a series on the earth sciences. It is intended for readers with some background in mathematics and physics, although for the reading of many sections this is not essential. In view of the rapid development of techniques, the author believes it best to emphasize physical principles rather than details of specific procedures. At the same time, several examples are given of geological problems in which gravity measurements have been of assistance.

Certain material has been placed in appendixes. This includes an outline of potential theory, which is essential for an appreciation of the subject but is not always convenient to locate in modern books. Finally, an introduction is given to the analysis of satellite orbits, but this is not developed in detail because the mathematical background required would be considerably more extensive than for the remainder of the book.

G. D. GARLAND

Edmonton, Alberta

vii

CHAPTER 1

Gravity, Geophysics, Geodesy and Geology

ONE of the most familiar facts about the earth is that a body released near it will fall with increasing velocity. The rate of increase of velocity is called the acceleration due to gravity, which Galileo showed to be the same for all bodies at a given point on earth. This is the most obvious example we have of a truly universal phenomenon, that of the mutual attraction of all masses. Newton formulated the principle of universal gravitation by deductions from Kepler's laws on the motions of the planets, showing that these laws were evidence of a force between each planet and the sun. The forces were shown to be proportional to the product of the masses involved, and inversely proportional to the square of the distance between them.

In the case of a body on the earth, the force of attraction is determined by the product of the earth's mass and the mass of the body, and the distance between the body and the earth's centre. If the earth were a uniform, non-rotating sphere, the force on a body at a given distance would be everywhere the same, and there would be a single constant value of the acceleration due to gravity. In fact, our earth is non-uniform, non-spherical, and rotating, and all of these facts contribute to variations in the acceleration over its surface.

Measurements and analyses of the variation in gravity (that is, the acceleration due to gravity) form a powerful branch of the science of geophysics, which is the investigation of the earth by the methods of physics. Those variations in gravity which are related to the departure of the earth from a spherical form are of particular interest in geodesy, the science of the earth's shape. Variations which reflect the non-uniform density of the earth can be used to infer the presence of structures beneath the surface, and are of interest to the geologist. The aim of the present book is to provide a general background for the appreciation of both applications of gravity measurements.

In geophysics, geodesy, and geology, there are other methods available for studying the shape and internal nature of the earth. Where possible, these methods are mentioned, and their relation to gravity studies are discussed. As one example, in the investigation of the earth's crust, there is a considerable advantage to combining results obtained from seismology with gravity measurements. On the other hand, limitations of space prevent a full discussion of other methods, especially in the case of geodesy, and it should not be thought that these methods are considered inferior.

Fundamental Concepts

Newton's law of gravitation for two particles of masses m_1 and m_2, separated by a distance r, is

$$F \propto \frac{m_1 m_2}{r^2}$$

or

$$F = G \frac{m_1 m_2}{r^2}, \tag{1.1}$$

where G is a constant. Here, F is the force on either mass, and is directed along the line joining the masses. The numerical value of the constant G was not determined in Newton's lifetime, being first measured in the laboratory by Cavendish in 1798. During the latter half of the eighteenth century, a number of attempts were made to measure G by determining the attraction of large features, such as mountains. None of these gave a quantitatively satisfactory result, although the work of Maskelyne (1774) was among the best. The Cavendish apparatus, which is well known, made use of the fact (Appendix 1) that the attraction between spheres is the same as that between massive particles at their centres, and provided for the measurement of this force by the determination of the torque acting on a suspended beam. The Cavendish experiment has been repeated a number of times with improved apparatus. We shall use, for numerical calculations, the value determined by Heyl (1930):

$$G = 6 \cdot 67 \times 10^{-8} \text{ c.g.s. units.} \tag{1.2}$$

To investigate the relation between the constant of gravitation G and the acceleration due to gravity g, we consider a mass m on the surface of the earth, of mass M_e. Then, neglecting effects of rotation and non-uniformity of shape and density, the force exerted on the mass is

$$F = G \frac{M_e m}{R^2}, \tag{1.3}$$

where R is the earth's radius.

If the mass were released, in a vacuum, it would accelerate toward the earth's centre with the acceleration g, where

$$g = F/m$$

$$= \frac{GM_e}{R^2}. \tag{1.4}$$

The determination of g and G is thus sufficient to yield a value for the mass of the earth, or equivalently, the mean density of the earth. During the eighteenth century, attempts to measure G were often considered to be experiments for the determination of mean density. The results of Cavendish showed that the mean density was about $5 \cdot 4$ g/cm^3, and this was one of the first proofs that the interior consists of material considerably denser than surface rock. Heyl's value for G corresponds to a value of mean density of $5 \cdot 52$ g/cm^3.

The value of g varies between 978 and 983 cm/sec^2 over the earth's surface. For investigations of the shape, or of internal structures, it is necessary to measure variations of $0 \cdot 001$ cm/sec^2 or less, and it is convenient to introduce a smaller unit of acceleration. We designate 1 cm/sec^2 as 1 gal (after Galileo), and 1×10^{-3} cm/sec^2 as 1 milligal (mgal).

Gravity Measurements and Reductions

As will be shown in the next chapter, the measurement of gravity in absolute terms to an accuracy of 1 mgal is a difficult problem. Fortunately, for both geodetic and geophysical purposes, the *variation* in g over the earth is the quantity usually required, and this can be measured more easily. The most direct application to geophysics of the absolute value is that given in equation (1.4), that is, the determination of the earth's mass. However, in other fields of physics, particularly in the setting up of standards of pressure, temperature, electric current, etc., the absolute value of g plays a very important role, and the geophysicist is often approached for assistance in providing a value to use in a given laboratory.

It has been mentioned that g varies over the earth for a number of reasons, and it might appear hopeless to separate these effects. However, it will be shown that we can treat, in turn, variations due to the shape of the sea-level surface, variations with height of the land, and variations due to concealed masses. For the study of the latter, the effect of the first two factors is removed from the observed values through a series of reductions. The quantities obtained are known as gravity anomalies, since they indicate the presence of anomalous conditions within the earth.

The Potential and Equipotential Surfaces

A mass in the presence of an attracting body has energy, by virtue of the attraction. This energy, known as potential energy, can be evaluated by considering the mass to have been brought from infinity, and calculating the work done on it in the process. This is done in detail in Appendix 1, where it is shown that the force of attraction can be obtained from the potential energy by differentiation. As the potential energy is a scalar quantity, in contrast to the vector force, it is often convenient to describe the gravitational field in terms of it, rather than in terms of the force. In particular, we shall make use of the potential energy of a unit mass (usually with a change of sign), which is known simply as the potential of the field.

A gravitational field can be represented by surfaces over which the potential is constant, known as equipotential surfaces. The force vectors are everywhere normal to these surfaces, so that there is no component of force along them. Thus, the surface of a liquid in a gravitational field coincides with an equipotential surface, and for this reason the potential is a quantity of great importance in the study of the shape of the sea-level surface of the earth.

Mathematical functions which represent the potential have a number of remarkable properties. The most important of these are derived in Appendix 1. Some knowledge of these is essential for an appreciation of the chapters on methods of analysis of the gravitational field, but not, to the same extent, for those chapters dealing with the results of gravity surveys.

CHAPTER 2

Gravity Measurements

THE measurement of the acceleration due to gravity provides an interesting example of the difficulties which may be met in the attempt to measure a relatively simple physical quantity. An acceleration involves only the fundamental dimensions of length and time, but, as indicated before, the measurements must be sufficiently precise to show extremely small variations in the quantity. To measure g precisely at some point on earth, in absolute units, without reference to any other point, is in fact an experiment requiring the greatest care. On the other hand, it is now relatively straightforward to measure the *differences* in g from place to place. In theory, therefore, an absolute measurement is needed at only one place, as the value of gravity at all other places could be determined with reference to it. However, it is desirable to have several absolute determinations, made with different types of apparatus, so that intercomparisons can be made, to indicate the precision achieved.

In this chapter, methods which have been used for absolute measurements will be described first, then instruments suitable for measuring differences in gravity will be discussed.

Absolute Measurements

Measurements of g in terms of length and time must be based on some physical equation which relates that quantity to times and distances which can be measured. The most direct approach is to time a freely-falling body over a known distance, but it is only recently that timing standards adequate for the precise measurement of short time intervals have become available. Almost all of the older determinations made use of some form of pendulum, as the expression for the period of a pendulum involves g, and the period can be measured by timing a large number of oscillations. The principles of both pendulum and falling-body experiments will be discussed before a detailed comparison of modern determinations is given.

Pendulum Measurements

The period of a simple pendulum swinging with vanishingly small amplitudes is

$$T = 2\pi\sqrt{\frac{l}{g}}, \tag{2.1}$$

where l is the length of the weightless, rigid, inextensible cord supporting the mass particle. If a simple pendulum could be

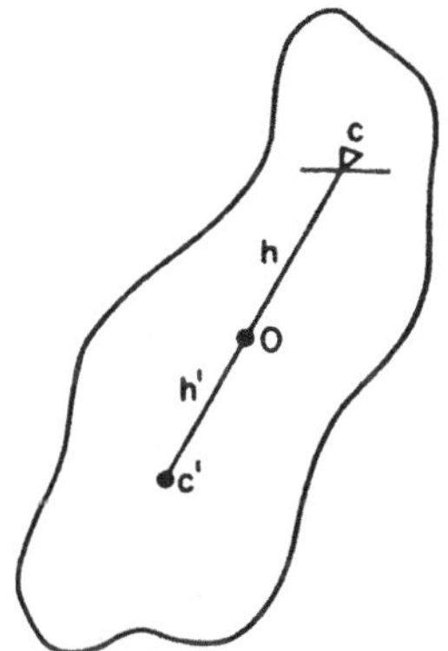

FIG. 2.1.   A physical pendulum.

constructed, g could be determined from measurements of T and l. However, any real body which can be made to swing on a pivot constitutes a physical pendulum (Fig. 2.1), for which the period of oscillation is

$$T = 2\pi\sqrt{\frac{I_c}{mgh}}. \tag{2.2}$$

Here, I_c is the moment of inertia of the pendulum about an axis through the pivot, m is its mass, and h is the distance from the pivot to the centre of mass. Since neither I_c nor h can be measured with precision, this expression is not suitable for the calculation of g as it stands. For every physical pendulum with a given point of support, there is an "equivalent simple pendulum" which has the same period T. The length l of the equivalent simple pendulum is given by

$$l = \frac{I_c}{mh}. \tag{2.3}$$

In terms of the moment of inertia about an axis through the mass centre, I_0,

$$l = \frac{I_0}{mh} + h \tag{2.4}$$

or

$$l - h = h' = \frac{I_0}{mh}. \tag{2.5}$$

The point C' in Fig. 2.1, at a distance l from the pivot C, is known as the centre of oscillation, and it is easy to show from the equations (2.4) and (2.5) that if the pendulum is swung about an axis through C', the same period T will be observed. In other words, if any two points C and C', on opposite sides of the mass centre, can be found, such that the periods are equal, the distance CC' gives the length of the equivalent simple pendulum, and g can be determined by equation (2.1).

The above principle was first used by Kater (1818) to measure the absolute value of g at a site in London. Kater constructed a pendulum consisting of a metal rod, with knife edges on which it could be swung fastened near each end. The two periods were made equal by adjusting a sliding bob on the bar. Measurements of the period with the bob in this position, and of the distance between knife edges, then allowed Kater to compute the value of g. The site of his observations has been identified, and it is known that his value is about 35 mgal high. In view of the simplicity of his apparatus, and of the limitations of his time standards, his value was surprisingly good.

Kater's method formed the basis of measurements made at Potsdam in the early years of this century, and more recently at Teddington and Washington. The measurements at Potsdam (Kühnen and Furtwängler, 1906) were carried out with six pendulums of different mass, and after a most careful series of experiments, designed to eliminate a number of sources of error, the value

$$g = 981 \cdot 274 \text{ gal}$$

was adopted for the site. This value remains the basis for the world gravity network, although it is now known to be in error by several milligals.

The measurement at Washington (Heyl and Cook, 1936) was made with a series of tubular pendulums constructed of fused silica. The pendulums had planes near each end, which could be rested on a fixed knife edge for the swings. To adjust the periods in the two positions to equality, one end of the pendulum was ground and polished. The observations were extremely tedious, as a series of swings with the pendulums in each position had to be made between each grinding operation.

The Teddington measurement (Clark, 1940) was made with a single metal pendulum, with a stem of I-section carrying bobs near each end. The knife edge, of hard steel, was fastened to the support, and the pendulum in swinging position rested on planes fastened to its stem.

The best available methods of measuring time and length were employed in the Washington and Teddington determinations, and these methods should have provided values of g accurate to 1 mgal. Measurements of the differences in gravity between Potsdam, Washington and Teddington, however, indicated a discrepancy of 5 mgal between the latter two, and an even greater difference with Potsdam. The explanation lies in a number of effects which produce systematic errors in pendulum observations. It is known that a swinging pendulum tends to drag the knife edge or support from side to side with it. The knife edge is thus not a fixed point, and the pendulum is actually rotating about a point above it, so that its effective length is greater than that measured. Kühnen and Furtwängler (1906) reasoned that the sway of the support would be proportional to the mass of the pendulum, and having made determinations with several pendulums of different mass, they extrapolated to zero mass. Later writers have questioned whether this extrapolation was justified. Other errors are introduced by the bending and stretching of the pendulum as it swings. The distance between knife edges or planes is, of course, corrected for the stretch of the pendulum when it is hanging in position, but the difficulty is that the periods are not the same as those of a perfectly rigid pendulum. Jeffreys (1948) has shown that the effect of bending is to lengthen the period, so that the measured value of g is too small if no correction is made for this effect. The correction can be estimated by computing the additional moment necessary to prevent bending, and determining the reduction of period which would be produced by this moment. In the case of the Teddington and Washington observations, corrections of this type amount to between 1 and 2 mgal.

Falling-body Methods

Qualitative experiments on falling bodies were made by Galileo, but it is only in recent years that quantitative measurements of g by this method have been possible. A body dropped from rest falls very nearly 5 m in the first second, and as 5 m is about the longest practical drop with which to work, a time of the order of 1 sec must be measured with an accuracy of one part in a million. Because of small initial accelerations which would be given to any body at the instant of release, distances and times cannot be measured from this instant, but must be measured between points on the path of fall.

If measurements of space and time could be made from the true origin of both, g could be calculated from the expression

$$g = 2S/t^2. \tag{2.6}$$

For measurements from an arbitrary origin, two values of

distance, S_1 and S_2, and the corresponding times t_1 and t_2, are required. Then g is given by

$$g = \frac{2\,(S_2 t_1 - S_1 t_2)}{t_1 t_2 (t_2 - t_1)}.\qquad (2.7)$$

This would suggest the necessity for three optical systems of known vertical separation, to locate the body at the origin of time and two succeeding times, but Volet (1952), who made the first

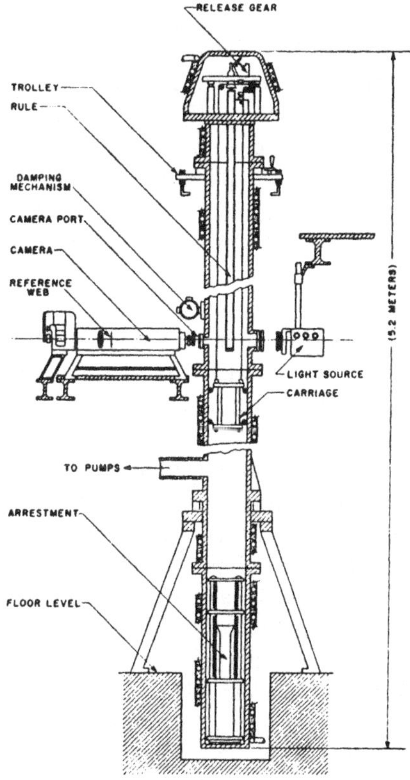

FIG. 2.2.   Absolute gravity apparatus used at the National Research Council, Ottawa. (After Preston-Thomas *et al.* Illustration courtesy of National Research Council.)

modern determination of g by this method, pointed out that if a graduated rule is dropped, only one optical system is required. Volet's type of apparatus was brought to a high state of refinement at the National Research Council, Ottawa (Preston-Thomas *et al.*, 1960), and a diagram of the arrangement used there is shown in Fig. 2.2. The rule dropped is of stainless steel, approximately 2 m long, and carrying three short glass scales, ruled at intervals of $0 \cdot 1$ mm. The scales were positioned so that as the rule dropped, they passed the axis of the camera at intervals of $0 \cdot 1$ sec. The rod was photographed at these times by means of a short-duration spark light source, which discharged ten times per second. An accuracy of 1 mgal in g was aimed for, which required the measurement of lengths to $0 \cdot 25 \times 10^{-3}$ mm, and times to 2×10^{-8} sec. The distance between the zeros of the glass scales was determined by direct comparisons with a standard metre, while the precise location of the rule at a given time was determined by scaling of the photographs, each of which showed a portion of a scale and a reference line. The standard of time was a 100 kc/s quartz crystal oscillator, from which a 10 c/s derived signal triggered the flash. Great care was taken to ensure that successive flashes had the same characteristics, and the crystal oscillator was compared with national time standards to give time in absolute units.

As the length of the rule was measured outside of the chamber, corrections were required for its expansion under the low pressure of the chamber, and for temperature changes. The greatest uncertainty was in the estimation of the effect of residual gas in the chamber. Drops were made at a series of pressures, down to a pressure of 1×10^{-5} mm of mercury. At these very low pressures the chief effect is the integrated force due to the collision of the rule with individual air molecules. The apparent value of g determined from the drops did indeed increase with decreasing pressure, but not at the rate predicted by theory. As most of the drops had been made at a pressure of 7×10^{-5} mm of mercury, there remained an uncertainty of about $0 \cdot 5$ mgal in the extrapolation to zero pressure.

It is characteristic of falling-body experiments that most of the error in the derived value of g is contributed by errors in length measurement, and because of the form of equation (2.7), the proportional error in the final value is several times the expected error of a measurement of length. This effect can be significantly reduced if the experiment is made symmetrical, that is, if an object is projected vertically upward, and timed as it rises and then falls past a fixed interval (Cook, 1957). If we let S be the length of the interval, and t_1 and t_2 the intervals of time between crossings of the lower and upper boundaries, respectively, of S, then g is determined from

$$g = \frac{8S}{t_1^2 - t_2^2}. \tag{2.8}$$

The relative error in g contributed by errors in length measurement is in this case the same as the relative error in S. A second important advantage of the symmetrical experiment is that forces on the object, proportional to its velocity, do not affect the times t_1 and t_2. The forces exerted by residual gas in the chamber are of this type, and there is, therefore, no need to extrapolate the observations to zero pressure.

In the apparatus which is being constructed for a redetermination of the absolute value of g at Teddington, the object to be projected upward is a glass ball (Fig. 2.3). As the ball passes between a pair of slits in each block, it acts as a lens, focusing light from one slit on the other. The light-sensitive detection circuit thus indicates the time of passage of the ball, while the distance interval between the blocks is simultaneously measured interferometrically, by comparison with the reference etalon. The distance between both top and bottom surfaces of the blocks can be determined, and uncertainties in the positions of the slits relative to these surfaces are eliminated by repeating the observations with the blocks inverted. It is noteworthy that the overall size of the apparatus is much less than that of unsymmetrical falling-body arrangements.

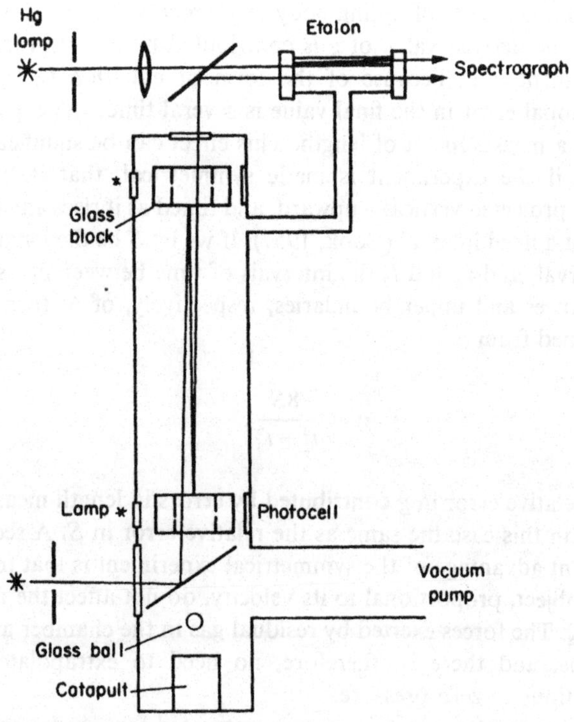

Fig. 2.3. Principle of absolute gravity method to be used at the National Physical Laboratory.

Comparison of Absolute Determinations

Relative measurements have now been made between the sites of the principal absolute determinations, and as these connections give the differences in g to an accuracy of better than 1 mgal they indicate the discrepancies between the absolute values themselves. The results are shown in Table 1.1.

It is apparent that the Potsdam value is approximately 14 mgal too high, but there has been a reluctance to adopt a new standard until one with an uncertainty of 1 mgal or less is available. As mentioned previously, an error in the absolute value does not

TABLE 2.1. ABSOLUTE MEASUREMENTS OF g.

Station	Institute	Absolute measurement as revised (gal)	Value on Potsdam system (gal)	Difference (mgal)
Sèvres	Bur. Int. Poids et Mesures	980·9280	980·9406	12·6
Teddington	Nat. Phys. Lab.	981·1832	981·1962	13·0
Ottawa	Nat. Res. Counc.	980·6132	980·6279	14·7
Washington	Nat. Bur. Stds.	980·0826	980·0990	16·4

affect conclusions about the earth based on the *variation* of gravity over the surface.

Relative Measurements

Pendulums

The use of the physical pendulum to measure differences in gravity, as distinct from the absolute value, dates back to the seventeenth century. Bouguer (1749) compared the period of a pendulum at Paris with the period at various places in South America, in connection with investigations on the shape of the earth. By the nineteenth century, the pendulum was in fairly general use as a relative gravity instrument.

The principle is very simple. Referring to equation (2.2), it is seen that if the same physical pendulum is swung, under identical conditions, at two places where the values of gravity are g_1 and g_2, the corresponding periods T_1 and T_2 will be related by the equation

$$\frac{T_1^2}{T_2^2} = \frac{g_2}{g_1}. \tag{2.9}$$

Differences in gravity can therefore be calculated by the equivalent equation

$$\Delta g = \frac{-2g_1}{T_1}(T_2 - T_1) + \frac{3g_1}{T_1^2}(T_2 - T_1)^2. \tag{2.10}$$

which shows that only the two periods T_1 and T_2 are required.

While the pendulum used in this way is a relative instrument, in that it measures differences in gravity, it has the great advantage of measuring this difference in absolute units, that is, in milligals. Other instruments, which are described later, measure the difference in arbitrary scale divisions, and must be calibrated.

The accuracy which can be obtained in relative pendulum measurements depends upon the accuracy of timing, the degree to which the conditions, such as temperature, are kept identical at the different sites, and the stability of the pendulum itself. Most pendulums which have been widely used for the purpose have equivalent lengths of $0 \cdot 25$ m and periods of approximately 1 sec. In this case, an error of 5×10^{-7} sec in the period corresponds to an error of 1 mgal in the gravity difference. With chronometers controlled by quartz crystal oscillators it is not difficult to achieve this accuracy in the period measurement with a swing lasting $\frac{1}{2}$ hr, but in the days of mechanical chronometers the period measurement was a major problem. The usual method of timing is to record, photographically or otherwise, signals from the pendulum and standard time marks produced by the chronometer. The record need not extend through the entire swing, as short records at the beginning and the end are sufficient. The period of the pendulum is always known approximately, and the correct integral number of swings between coincidences of time marks and pendulum signals can be calculated.

Ideally, pendulums should be swung in an evacuated case, at constant temperature, with the same amplitude. In practice, corrections can be applied if the pressure, temperature and amplitude are not identical at all stations. There is a very considerable advantage to swinging two pendulums, simultaneously but in opposite phase, in the same case. The forces exerted by the pendulums on the case annul each other, if the pendulums are in the same plane, and the sway of the case is eliminated. Also, the mean period of the two pendulums is, to the first order, independent of horizontal ground accelerations which are serious in many locations. Differences in gravity can be calculated using this mean period.

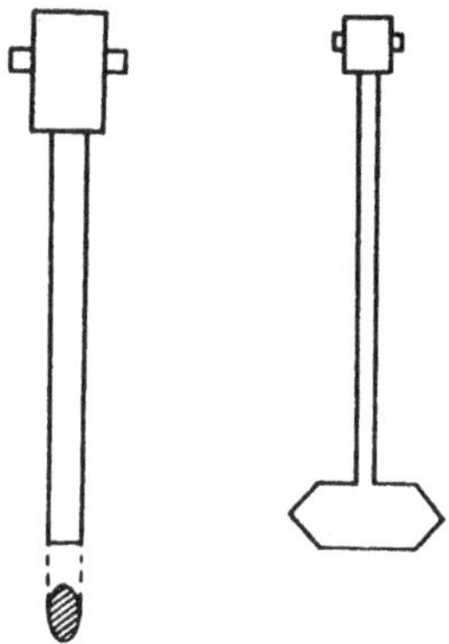

FIG. 2.4. Form of Gulf quartz (left) and Cambridge invar pendulums.

The shapes of the pendulums used in the Cambridge (Jackson, 1961) and Gulf (Gay, 1940) apparatus are shown in Fig. 2.4. The former are of invar, an alloy of iron and nickel with small coefficient of thermal expansion, while the latter are of fused quartz. Both of these pendulums have performed very well, although care is necessary to avoid the influence of the earth's magnetic field on the Cambridge pendulums. The shape and location of the knife edges of the Gulf pendulums were chosen to minimize changes in the period due to wear of these edges. Referring to equation (2.4), and writing mk^2 for I_0, where k is the radius of gyration about an axis through the mass centre,

$$l = \frac{k^2}{h} + h. \qquad (2.11)$$

$$\therefore \frac{dl}{dh} = 0 \quad \text{when} \quad h = k.$$

That is, when the knife edge is a distance equal to k from the mass centre of the pendulum, the period is independent of small changes in this distance. This condition cannot be realized when the pendulum consists of a thin stem with massive bob.

Changes in the period due to shock or other physical damage

can only be reduced by the use of great care during transport and observations. The period is always measured at the base station both before and after the observations at other stations, so that any change in properties can be detected. Since the chief purpose of pendulum measurements now is to provide calibration bases for other gravity measuring instruments, the emphasis is currently on the establishment of a relatively few stations, of high accuracy, distributed over the earth.

Static Gravity Meters

In 1849 Sir John Herschel pointed out that it should be possible to measure differences in gravity with a mass hung on a spring. The tension in the spring produced by the weight of a mass m is mg, and if the instrument is taken to places where g is different, the length of the spring should change. It is not difficult to see that refinements are necessary if differences in g are to be measured with precision, and, in fact, Herschel himself did not succeed in constructing a useful instrument. Apart from disturbing influences, such as changes in the temperature of the spring, there is the fact that the length of the spring in this simple instrument depends on the total value of g, and direct measurement of changes in length of the order of 1 part in 10^6 are difficult in a portable instrument. The solution lies in either converting the displacement of the system to some quantity, such as a rotation, which is easier to measure, or using a system in which a small difference in gravity produces relatively large displacements. Many instruments capable of measuring differences to 0·1 or even 0·01 mgal were developed in the years between 1930 and 1950. Examples of types using optical or electrical magnification are shown in Fig. 2.5.

The Gulf gravimeter, shown here, makes use of the rotation of a spring, which accompanies the spring's extension. If the cross-section of the wire is correctly chosen, the rotation can be made very large for a given extension. The other instrument illustrated in Fig. 2.5, the Boliden, makes use of the fact that the capacitance of a parallel-plate condenser varies with the separation of the

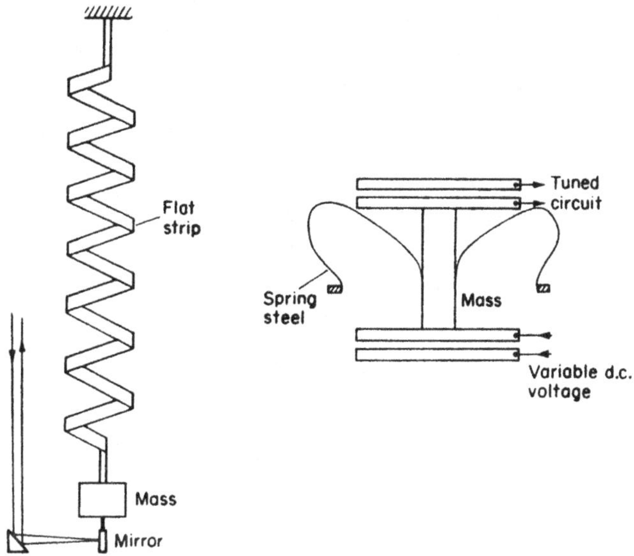

FIG. 2.5. Principles of Gulf (left) and Boliden gravimeters.

plates, and small changes in capacitance in a tuned circuit can be easily detected.

In other instruments, an assembly resembling a vertical seismograph is used (Fig. 2.6). Let us suppose the spring supporting the mass is wound in such a way that its tension is proportional to its

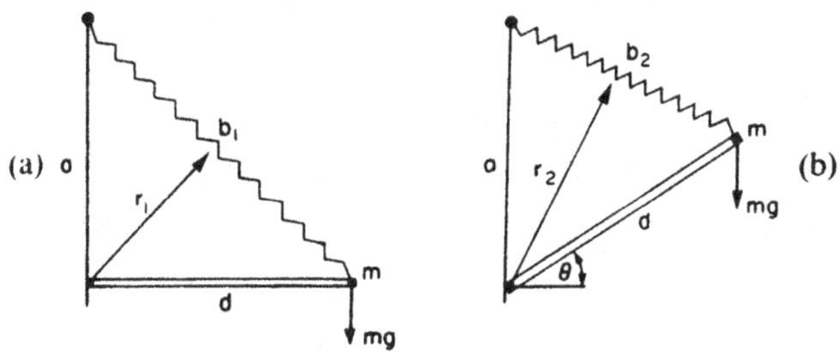

FIG. 2.6. A mass supported by a zero-length spring.

physical length. If there were no tension in it, the length would be zero, and for this reason it is known as a zero-length spring (Lacoste, 1934). Of course, the spring cannot exist with zero length. If it were lying on a table, it would have its minimum physical length, and it would be in a state of tension because of forces exerted between its coils. The simplest way to picture such a spring is to imagine a normal coil spring to be turned inside out. If such a spring, with proportionality constant K, supports the mass when the beam is horizontal, it will also support it in any other position. Equating moments about the pivot, in case (a) of Fig. 2.6, we have

$$mgd = Kr_1b_1$$
$$= Kad. \tag{2.12}$$

When the beam is displaced as in case (b), the moment due to the weight becomes $mgd \sin \theta$, while that due to the spring becomes

$$Kr_2b_2 = Kad \sin \theta. \tag{2.13}$$

If equation (2.12) is satisfied, equilibrium is maintained for all values of θ.

A suspension employing a spring of near zero-length is therefore subject to a very small restoring force, and is said to be astatized. If it is balanced for one value of g, and taken to a place where g is different, the displacement of the system is relatively large. This principle is used in the Worden gravimeter, shown in Fig. 2.7. All parts of the system, except the metal strip for temperature compensation, are of fused quartz. The mass is always brought to a standard position, as indicated by the pointer viewed in the eyepiece, by adjusting the dial connected to the measuring spring. If the instrument is taken to a place where g is very different, so that the change is beyond the range of the measuring spring, the reset spring must be altered. In normal use, for relatively local surveys, the reset adjustment is used only to bring the instrument on scale, and all measurements are made with the dial connected to the measuring spring. However, a model has been constructed, for geodetic purposes, with a dial fitted to the reset

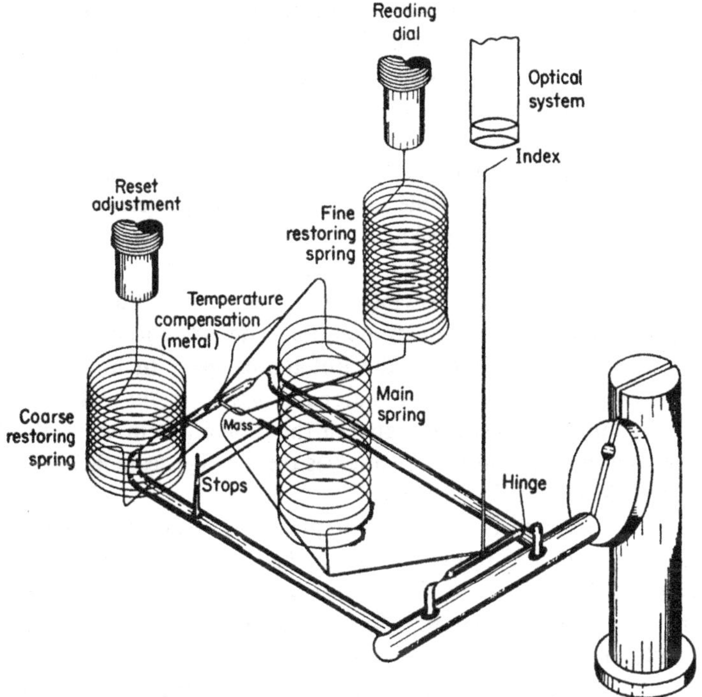

Fɪɢ. 2.7. The Worden gravimeter.

spring (Woollard, 1950), so that very large differences in g can be measured, although with lower precision.

Most gravimeters are maintained thermostatically at constant temperature. Temperature changes on the Worden system are reduced by mounting it inside a vacuum flask, but the temperature compensation strip is adjusted so that its differential expansion keeps the mass in position when temperature does change.

It is apparent from the above descriptions that gravimeters may be expected to have two common characteristics. First, differences in g will be obtained initially in terms of arbitrary scale divisions, and secondly, any change in the elastic properties will cause the reading at one place to change or "drift" with time. On the other hand, these instruments are very portable, weighing as little as 5 lb, they can be read very rapidly, and they can detect smaller differences in g than any other instrument. The great increase, in recent years, in the number of gravity stations on the earth is due to their development.

As the response of the gravimeter system is not, in general, linear over the total possible range of gravity, it is essential that the instrument be calibrated against known values of g in the same part of the range as that in which it is to be used. In other words, the gravimeter is an effective instrument for interpolation between control points, but not for extrapolation. The only effective way to calibrate it is to determine the differences in g, in instrument divisions, between pendulum stations at which gravity is known. A least squares adjustment between the differences in scale divisions and those in milligals then gives the average scale factor for that particular gravity interval.

If an instrument is set up at one place, and read from time to time, the reading is found to change. This change will be contributed partly by drift in the instrument, and partly by the real time-changes in g, due to tidal forces. For accurate determinations of differences in gravity, between stations, readings must be repeated at some base point, to permit correction for the change in reading. If a survey covering a large area is to be undertaken, a number of points are usually chosen as bases, and these are connected with great care. Figure 2.8 indicates the sequence of

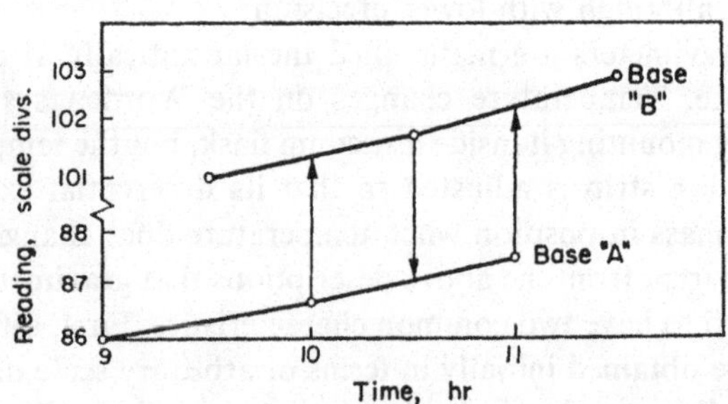

FIG. 2.8.   Drift curves constructed by repeat readings of a gravimeter

readings at two points, and the typical appearance of "drift curves" plotted from the repeat readings at each. The best estimate of the difference in reading between the points is the mean distance between the curves. It will be noted that if readings are repeated each hour or so, the resultant drift and tidal effect can be treated as a straight-line change of reading with time. Other stations of the survey would normally be observed on single loops, beginning and ending at any base point, and linear drift curves would be drawn for each circuit, based on these two readings.

Measurements from a Moving Vehicle

Our knowledge of the earth's gravitational field would be severely restricted if gravity measurements were possible only on land. The fact that so much of the surface is covered by water makes a seaborne instrument essential, and in recent years there has been interest in the possibility of an airborne gravimeter.

The difficulty in any measurement of g from a moving vehicle arises from accelerations of the support. As g itself is an acceleration, it is impossible in principle to separate it completely from these accelerations. It remains to be seen what degree of approximation is possible.

Vening Meinesz (1929), in his pioneer work on gravity observations at sea, investigated the first-order effects of these accelerations on a swinging pendulum. We consider first the equation of motion of a pendulum, of equivalent length l, oscillating on an undisturbed support.

The equation is

$$\ddot{\theta}_1 + \frac{g}{l}\theta_1 = 0. \tag{2.14}$$

where θ_1 is the angular displacement from the vertical. If the support is subject to a horizontal acceleration $\ddot{y}$, in the plane of oscillation, the equation becomes

$$\ddot{\theta}_1 + \frac{g}{l}\theta_1 = -\frac{\ddot{y}}{l} \tag{2.15}$$

and the period of the motion is no longer simply related to g. Vening Meinesz noted that if a second, identical, pendulum were swung on the same support, its equation would be

$$\ddot{\theta}_2 + \frac{g}{l}\theta_2 = -\frac{\ddot{y}}{l} \qquad (2.16)$$

and the difference between equations (2.15) and (2.16) gives

$$(\ddot{\theta}_1 - \ddot{\theta}_2) + \frac{g}{l}(\theta_1 - \theta_2) = 0. \qquad (2.17)$$

In other words, a "fictitious pendulum" oscillating with the difference in amplitude $(\theta_1 - \theta_2)$ behaves as a pendulum swung on a disturbance-free support, and differences in g can be deduced from its period. Vening Meinesz devised an arrangement of three pendulums, in which the motions of two fictitious pendulums were directly recorded optically, by successive reflections of a beam of light from mirrors on two pendulums of a pair.

Vertical accelerations of the support cannot be separated from gravity, and can only be reduced if the observations extend over a period of time, and if the accelerations have a small mean value. If $\ddot{z}$ is the vertical acceleration, the mean is

$$\bar{\ddot{z}} = \frac{1}{t}\int_0^t \ddot{z}\,dt = \frac{1}{t}\left[\dot{z}_t - \dot{z}_0\right]. \qquad (2.18)$$

The mean thus depends on the initial and final vertical velocity of the support, and on the time interval, t. In Vening Meinesz's measurements, which were made in submarines submerged below the depth of wave action, $\ddot{z}$ became effectively 0 when t was made 30 min or more. However, if some type of gravimeter is to be used in a surface ship or aircraft, the vertical velocities might well have to be measured, so that equation (2.18) could be evaluated. For example, imagine that the vertical velocity of an aircraft could be measured to 1 cm/sec. For equation (2.18) to be evaluated to 1 mgal, the observation would have to be averaged over 16 min. During this time, the aircraft may have covered 100 miles over the earth's surface. Airborne measurements would thus appear to be of the most use in providing a generalized picture of the gravitational field, for geodetic purposes, rather than detailed measurements.

When the instrument used responds to the total gravity vector there is an important second-order effect in measurements from a moving support. This effect, first noted by Browne (1937), is not included in any of the above considerations. If the instrument support is subject to horizontal accelerations $\ddot{x}$ and $\ddot{y}$, and if these disturbances have periods much longer than the natural period of the instrument, the quantity measured will be the resultant vector,

$$g\left(1 + \frac{\ddot{x}^2 + \ddot{y}^2)}{2g^2}\right.$$

As the second term involves only squares, the expression is systematically greater than g, and the error is not reduced by extending the period of observations. Some independent measurement of the horizontal accelerations is necessary, to allow the correction to be calculated. Alternatively, a gravimeter, sensitive to the component of gravity in a given direction, may be used in such a way that it is maintained in a fixed orientation by a gyroscopically-maintained control system (Lacoste and Harrison, 1961). Gravimeters which have been used on surface ships include the Graf (Graf and Schulze, 1961), and the Lacoste–Romberg (Lacoste, 1959). In the Graf instrument, which is normally used with gyroscopic control, a continuous recording is made of the displacement of a very heavily damped beam. The position of the mean of this curve indicates the relative value of gravity. For the Lacoste–Romberg instrument, a damped gimbal mounting is used. The suspension thus responds to the total gravity vector, and the Browne correction is required. However, horizontal accelerometers are used to measure the accelerations, and the correction is computed and applied to the reading automatically. The tension in the beam suspension as a function of time is

B

recorded, in digital form, and the mean value over an interval time is taken as the relative value of gravity. Satisfactory tests of the Lacoste–Romberg instrument in an aircraft have been reported (Nettleton, Lacoste and Harrison, 1960).

Finally, there is the fact that, independent of accelerations, the value of g for an observer moving relative to the earth is not the same as that for an observer rotating with the earth. The Eötvös effect, as it is known, arises from the change in centripetal acceleration. A point on the earth, at distance r from the axis of rotation, has a centripetal acceleration a toward that axis, of $r\omega^2$, where the angular velocity of rotation of the earth. The observed value of g is reduced by this centripetal acceleration. If, because of an east–west velocity v, relative to the earth, the observer's angular velocity changes by $d\omega$, then

$$da = 2r\omega d\omega = 2\omega v. \tag{2.19}$$

A component of this is in the direction of gravity. If φ is the geocentric latitude of the observer, the apparent change in gravity dg, is

$$\left. \begin{aligned} dg &= da\cos\varphi \\ &= 2\omega v\cos\varphi \\ &= 7\cdot 5v\cos\varphi \end{aligned} \right\}, \tag{2.20}$$

where dg is in milligals and v in miles per hour. Gravity is increased when the observer moves from east to west over the earth. Equation (2.20) shows that, in middle latitudes, a component of velocity of even 1 m.p.h. east to west corresponds to a change in g of 5 mgal. The magnitude and direction of the velocity of the ship or aircraft from which g is to be measured must be accurately known, if a precision approaching 1 mgal is to be achieved. In the case of aircraft speeds, a similar but smaller effect due to north–south motion of the observer may also be significant.

Gravity Networks

Regardless of the type of instrument used to measure differences in g, or of the care taken to remove the effects of drift, errors will accumulate in any series of measurements. This can be illustrated if differences are measured around the sides of a closed figure; usually there is found to be a closure error when the differences are added. The obvious conclusion is that if gravity is to be measured over a large area, the work should be laid out in a series of interlocking closed circuits. Measured differences may then be adjusted by least squares to give minimum closure errors for all circuits, in much the same way as is done with geodetic level nets. If a number of pendulum stations are included in the network, the adjustment of the differences, in instrument divisions, will also yield the scale constant of the gravimeter.

Ideally, the surface of the earth should be covered by a single network of closed circuits, for which a uniform adjustment could be made. Difficulties have arisen because of the number of instruments which have been used for inter-continental measurements, and because different connections must be given very different weights. The International Association of Geodesy has proposed a list of first-order gravity stations distributed over the earth, and has urged that priority be given to establishing connections between them. When these have been made, a rigorous world-wide adjustment will be possible.

CHAPTER 3

The Shape of the Sea-level Surface

Introduction

The value of g varies over the earth's surface for a number of reasons. In this chapter, particular attention will be given to those variations which are related to the earth's shape. The departure of the form of the sea-level surface from a sphere leads to variations in the value of g, even at sea level; conversely, a study of these variations is of great assistance in determining these departures. In practice, gravity is only occasionally measured at sea level, for most stations on land are at some height above it. The measured values at these stations can be corrected to sea level by procedures which will be introduced later. For the present, we simply assume that sufficient measurements of gravity at sea level are available for the world-wide variation to be investigated.

The study of the shape of the sea-level surface of the earth is intimately related to many of the major problems of geodesy. It is appropriate here to outline the chief aims and methods of that science.

The Aims of Geodesy

A knowledge of the shape of the earth, and of the location of the land masses on it, is essential for mapping. The geodesist provides the basic control for the location of points on the earth in terms of latitude, longitude and height above sea level. If these points are to be located in space, the form of the sea-level surface must be known.

It has already been pointed out that sea level, undisturbed by winds or tides, is an equipotential surface of the earth's gravitational field. Where there are local variations in g, due to internal density anomalies, this surface is distorted. Most of these distortions are rather limited in extent over the surface, and all of them are of very small amplitude compared to the earth's radius. If the earth were viewed from a considerable distance, say from the moon, they would not be visible, and the earth would appear as a slightly flattened sphere, or spheroid. It is convenient, therefore, to adopt as a first approximation to the sea-level surface of the earth a spheroid. If the warpings of the actual equipotential or sea-level surface from this spheroid can be determined, the first part of the geodesist's problem is solved. Measurements of gravity play an important role in defining both the spheroid and the actual equipotential surface. Clairaut (1743) showed the relationship between the uniform variation of sea-level gravity from equator to poles and the flattening of the spheroid, while Stokes (1849) proved that a knowledge of more local irregularities in the gravitational field could be used to determine the warpings of the sea-level surface.

Sea level can be measured only over the oceans, but the equipotential surface corresponding to it is a complete, closed surface, known as the geoid. The geoid under the continents can be thought of as the surface defined by water level in narrow canals cut through the land masses (Fig. 3.1). The land surface of the continents is defined by its "height above sea level", and this is indeed

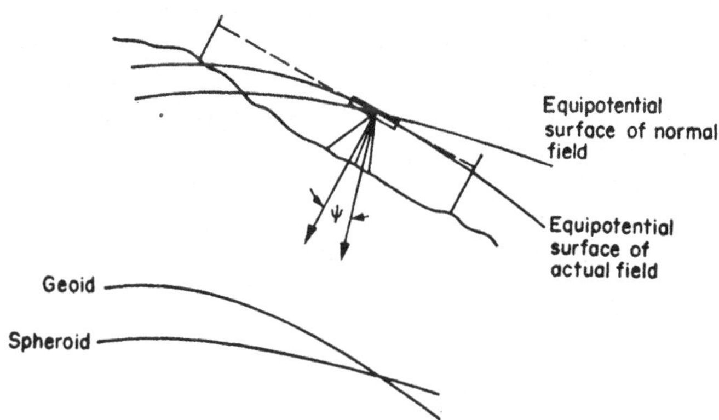

FIG. 3.1.   A geodetic level in the earth's gravitational field.

height above the geoid. Normally, heights are determined by extending lines of levels from the sea.

At each place where the spirit level is set up, the optic axis is placed in a plane tangential to the equipotential surface through that point (Fig. 3.1). This surface is very nearly parallel to the geoid, and the surveyor's readings of his levelling rods therefore give differences in height above the geoid, if we neglect the small effect due to non-parallelism of the surface equipotential surface and the geoid. The important fact at this stage is that points on the earth's surface are located with reference to the geoid. In fact, a second way of defining the geoid beneath the continents is as that surface which is everywhere at a depth equal to the measured height below the land surface.†

The location of points on the earth in latitude and longitude is complicated by variations in the direction of gravity. Since the direction of the total vector is determined by the direction of any local anomalous attraction as well as by the direction of the earth's normal attraction, it is variable from place to place. It is, of course, normal to the geoid (Fig. 3.1). This is the direction indicated by the plumb line at any point, and is the local vertical. Because of the effect of local influences on the vertical, position on earth cannot be determined with precision by any method which involves the measurement of a direction with reference to it. For example, astronomical determination of latitude and longitude makes use of the elevation of a circumpolar star, and the time of transit of a star across the meridian. Both of these observations are made with reference to the local vertical, and the position determined contains any anomaly which may be present in this direction.

Geodesists, in practice, locate position by triangulation, which involves chiefly the measurement of horizontal angles, and the length of a few base lines. Relative positions of points are thus determined with great precision, independent of local anomalies

† The two definitions that have been given for the geoid are not, in fact, identical, and the geodesist introduces other terms at this point. As the differences between the two surfaces are slight, we believe that the physical principles can be best appreciated if the distinction is not made here.

in the plumb-line direction. However, the net of triangulation must be started somewhere, and the position of the starting point must be assumed. Normally, the astronomical position of this point is adopted as correct; in other words, the direction of the vertical is taken to be normal there. If there is in fact a deflection of the vertical from its normal direction, the entire net of triangulation will contain this error. The land surface of the earth has by now been covered by a relatively few nets of very large extent over the various continents. In general, these nets have their own starting points, and there is the possibility of consistent discrepancies between them if the directions of the vertical at these points are not investigated. Stokes' result that the departure of the geoid from the spheroid could be determined from gravity anomalies showed also that the slope of the geoid, equivalent to the deflection of the vertical, could be computed. The determination of the geoid through gravity measurements thus plays an important role in the geodetic operations of both position and height location for points on the earth's surface.

The importance of an accurate knowledge of the vertical direction is not limited to mapping problems on the earth. It is vital to the accurate launching of space vehicles, if these are launched with reference to the local vertical direction. This is very often the case, as the launching tower or silo is aligned by plumb bob, and use may be made of inertial guidance systems, vertically aligned, in the vehicle itself.

The Spheroid

The spheroid is a mathematical figure which represents the actual sea-level surface with all local irregularities removed. It would, in fact, be the sea-level surface of an earth which had no lateral variations in density, only a uniform density variation from centre to surface. On such an earth, sea-level gravity would vary smoothly from the equator to the poles, and the spheroid would be an equipotential surface of this gravitational field.

In the choice of the exact form of the spheroid, attention must be given to theoretical considerations, direct measurement of meridian arcs, and the analysis of gravity variations. The external forms of a rotating, fluid mass of uniform density, and of density increasing toward the centre, can be calculated. As the density of the earth is known to increase with depth, the spheroid is chosen to be consistent with the latter case. By measurements of the length on the surface corresponding to a degree of latitude, as a function of latitude, it became known by the eighteenth century that the spheroid was oblate, with a flattening (Fig. 3.2) of approxi-

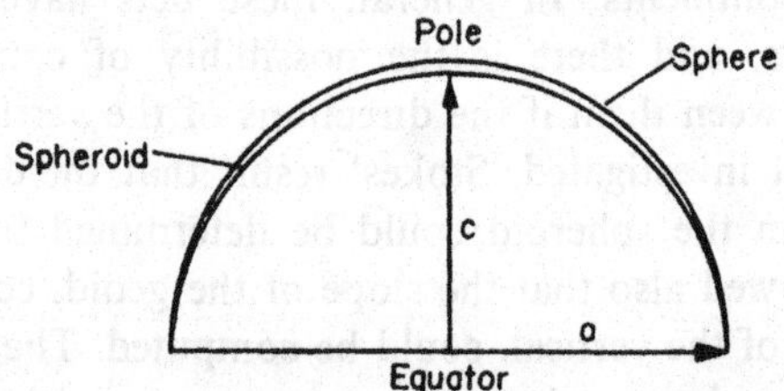

FIG. 3.2. Flattening of the spheroid $[(a-c)/a]$.

mately 1/300. This method yields slightly different values for the flattening when it is performed along different meridians, and the great advantage of determining the flattening from the gravitational field, as suggested by Clairaut (1743), is that the best mean value for the earth is obtained.

A mass of fluid under rotation assumes a form such that its external form is an equipotential of its own attraction and the potential of the centripetal acceleration. If the density is uniform, the form may be that of an ellipsoid of revolution (Appendix 2).

If the density of the fluid increases toward the centre, the external form is no longer that of an ellipsoid of revolution (Darwin, 1910). The surface is depressed below the ellipsoid, the amplitude of depression increasing uniformly from the equator and poles to a maximum at latitude 45°. In the case of the earth, the mean density is known to increase from approximately $3 \cdot 0$ g/cm³ at the surface to 11 or 12 g/cm³ at the centre, with a considerable proportion of the increase taking place at the core boundary. Darwin showed that the form of the surface is rather insensitive to the precise nature of the density variation, and that any reasonable assumption gives a maximum depression below an ellipsoid of about 3 m.

Clairaut's Theorem

Clairaut (1743) deduced the relation between the variation of gravity from equator to poles, and the flattening of the spheroid. His development was valid only to terms of the first order in the flattening, and to this order there is no distinction between the spheroid and an ellipsoid of revolution. The following development is based on Helmert's (1884) treatment of the first-order problem.

The spheroid is taken as the equipotential surface which surrounds all the mass of the earth. (The effect of mass above sea level will be discussed later.) The origin O is at the mass centre of the earth, and the z-axis of Cartesian coordinates is the axis of rotation. Primed coordinates refer to a point of observation P, unprimed ones to a general point in the body. The potential of the gravitational field at a point on or outside of the surface can be written

$$U = G \int \frac{\mathrm{d}m}{\varrho} + \tfrac{1}{2}(x'^2 + y'^2)\omega^2. \tag{3.1}$$

The second term is the potential of the centripetal force on a unit mass. The term $1/\varrho$ appearing in the integral can be expanded in a series of Legendre polynomials (Appendix 1):

$$\frac{1}{\varrho} = \frac{1}{r'}\left[1 + \frac{r}{r_1}P_1(\cos\gamma) + \left(\frac{r}{r'}\right)^2 P_2(\cos\gamma) + \ldots\right], \tag{3.2}$$

where
$$P_1(\cos\gamma) = \cos\gamma,$$

$$P_2(\cos\gamma) = -\tfrac{1}{2} + \tfrac{3}{2}\cos^2\gamma, \text{ etc.}$$

Now $\cos\gamma$ can be expressed in terms of the Cartesian coordinates of P and Q:

$$\cos\gamma = \frac{xx' + yy' + zz'}{rr'} \tag{3.3}$$

and also in terms of their geocentric latitude and longitude:

$$\cos\gamma = \cos\varphi\cos\varphi'\cos(\lambda-\lambda')+\sin\varphi\sin\varphi'.$$

This gives

$$P_2(\cos\gamma) =$$

$$\tfrac{9}{4}(\sin^2\varphi-\tfrac{1}{3})(\sin^2\varphi'-\tfrac{1}{3})+3\sin\varphi\cos\varphi\sin\varphi'\cos\varphi'\cos(\lambda-\lambda')$$
$$+\tfrac{3}{4}\cos^2\varphi\cos^2\varphi'\cos2(\lambda-\lambda'). \quad (3.4)$$

The expression for U becomes

$$U = \frac{G}{r'}\left[\int dm +\frac{1}{r'}\int P_1(\cos\gamma)r dm +\frac{1}{r'^2}\int P_2(\cos\gamma)r^2 dm +\dots\right]$$
$$+\tfrac{1}{2}(x'^2+y'^2)\omega^2 \quad (3.5)$$

and all terms above $P_2(\cos\gamma)$ will be neglected.

The first integral $= M$, the total mass within the spheroid. When the expression for $P_1(\cos\gamma)$ is substituted in the second, the integral is found to consist of terms of the form $\int x dm$, which are zero in consequence of the selection of the mass centre as origin.

The third integral is given by

$$\int P_2(\cos\gamma)r^2 dm = \tfrac{3}{2}(\sin^2\varphi'-\tfrac{1}{3})\int\left(z^2-\frac{x^2+y^2}{2}\right)dm$$
$$+3\sin\varphi'\cos\varphi'\left[\cos\lambda'\int xz dm+\sin\lambda'\int yz dm\right]$$
$$+\tfrac{3}{4}\cos^2\varphi'\left[\cos2\lambda'\int(x^2-y^2)dm+\sin2\lambda'\int 2xy dm\right] \quad (3.6)$$

The terms on the right of equation (3.6) obviously involve moment and products of inertia. In any solid body, principal axes of inertia can always be found so that the products of inertia vanish. The axis of z must be the principal axis of greatest moment of inertia, as it is the rotation axis. Hence, it is only necessary to choose x and y so that they coincide with principal axes.

Then

$$\int P_2(\cos\gamma)r^2 dm$$

$$= \tfrac{3}{2}\left(\frac{A+B}{2}-C\right)(\sin^2\varphi' -\tfrac{1}{3})+\tfrac{3}{4}(B-A)\cos^2\varphi'\cos2\lambda', \quad (3.7)$$

where A, B and C are moments of inertia about x, y and z respectively. The potential may now be written

$$U = \frac{GM}{r}\left[1+\frac{K}{2r^2}(1-3\sin^2\varphi)+\frac{3(B-A)}{4Mr^2}\cos^2\varphi\cos2\lambda\right.$$
$$\left. +\frac{\omega^2 r^3}{2MG}\cos^2\varphi\right], \qquad (3.8)$$

where $K = \left(C-\dfrac{A+B}{2}\right)\Big/M$, and primes are no longer required on the coordinates of P.

The potential contains a longitude term only if $B \neq A$. For the first-order theory, the two equatorial moments of inertia are taken to be equal, and the potential is further simplified to

$$U = \frac{MG}{r}\left[1+\frac{K}{2r^2}(1-3\sin^2\varphi)+\frac{\omega^2 r^3}{2MG}\cos^2\varphi\right]. \qquad (3.9)$$

Equation (3.9) may first be used to solve for r in terms of φ, if P is constrained to lie on the equipotential surface $U = U_0$. In this case

$$r = \frac{MG}{U_0}\left[1+\frac{K}{2a^2}(1-3\sin^2\varphi)+\frac{\omega^2 a^3}{2MG}\cos^2\varphi\right]. \qquad (3.10)$$

The second and third terms on the right are small compared to unity, and the equatorial radius a may be substituted for the variable r. The term $\omega^2 a^3/MG$, which will henceforth be designated m, has a simple physical significance:

$$m = \frac{a\omega^2}{MG/a^2} = \frac{\text{centripetal acceleration at equator}}{\text{attraction at equator}}$$

which is of order $1/300$.

Equation (3.10) may be rearranged to give

$$r = \frac{MG}{U_0}\left(1+\frac{K}{2a^2}+\frac{m}{2}\right)\left[1-\left(\frac{3K}{2a^2}+\frac{m}{2}\right)\sin^2\varphi\right] \quad (3.11)$$

which is of the form $\quad r = a(1-f\sin^2\varphi)$

where the flattening f of the spheroid is given by

$$f = \frac{3K}{2a^2}+\frac{m}{2}. \tag{3.12}$$

Returning to equation (3.9), we can obtain g by differentiation. To the first order

$$g = -\frac{\partial U}{\partial r}$$

(the differentiation should actually be along the normal to the equipotential surface), which gives

$$g = \frac{MG}{r^2}\left[1+\frac{3K}{2r^2}(1-3\sin^2\varphi)+m\cos^2\varphi\right]. \tag{3.13}$$

Substituting for r from equation (3.10), we obtain g on the spheroid, designated γ_0, as

$$\gamma_0 = \frac{U_0^2}{GM}\left(1+\frac{K}{2a^2}-2m\right)\left[1+\left(2m-\frac{3K}{2a^2}\right)\sin^2\varphi\right]. \tag{3.14}$$

This has the form

$$\gamma_0 = \gamma_e(1+B_2\sin^2\varphi),$$

where $\qquad\qquad B_2 = 2m-\dfrac{3K}{2a^2} \qquad\qquad (3.15)$

(The quantity φ, which was introduced as geocentric latitude, may be taken as geographic latitude in the first-order theory.)

Comparison of equations (3.12) and (3.15) shows that

$$B_2 = \tfrac{5}{2}m-f, \tag{3.16}$$

which is Clairaut's result.

Clairaut's equation cannot be used as it stands for geodetic purposes, because of the neglect of terms of order f^2. Darwin (1910) and many later workers (Gulatee, 1940) carried through a similar analysis without neglect of these terms. Let sea-level gravity be fitted to an equation of the form

$$\gamma_0 = (1 + B_2 \sin^2\varphi - B_4 \sin^2 2\varphi), \qquad (3.17)$$

where φ is the geographic latitude.

The coefficient B_2, correct to order f^2, is

$$B_2 = \tfrac{5}{2} m - f - \tfrac{17}{14} mf - \tfrac{2}{7} \chi, \qquad (3.18)$$

where m and f have their previous meanings, and χ is a constant which is determined by the internal density distribution. Darwin (1910) showed that any reasonable assumption for the variation of density with depth led to a value for χ between -175×10^{-8} and -205×10^{-8}. The magnitude of χ is a measure of the departure of the spheroid from an ellipsoid.

In terms of the same quantities,

$$B_4 = \tfrac{1}{8} (f^2 - 5mf + 6\chi). \qquad (3.19)$$

A quick calculation shows that B_4 is of the order of 10^{-5}, and therefore much smaller than B_2. It represents a variation in g with an amplitude of something under 10 mgal, in the form of a smooth variation from zero at the equator and poles to a minimum at latitude 45°. Up to the present time, it has not been possible to deduce the value of the coefficient B_4 from the observed values of gravity. If the value of 200×10^{-8} for χ is adopted, $B_4 = 7 \times 10^{-6}$. If the spheroid is taken to be an ellipsoid of revolution, $\chi = 0$, and $B_4 = 5 \cdot 9 \times 10^{-6}$. Both of these values can be found in formulae for γ_0 which have been suggested at different times.

The International Formula, adopted by the International Union of Geodesy and Geophysics in 1924, is

$$\gamma_0 = 978 \cdot 049(1 + 0 \cdot 0052884 \sin^2\varphi - 0 \cdot 0000059 \sin^2 2\varphi). \quad (3.20)$$

Here, $978 \cdot 049$ is the derived value of equatorial sea-level gravity, on the Potsdam standard. The value adopted for B_4 shows that a true ellipsoid has been taken as the reference surface, and the value of B_1 corresponds to a flattening of 1/297.

More recent gravity formulae, based on the greater number of observations available, include Jeffreys' (1948):

$$\gamma_0 = 978 \cdot 0373(1 + 0 \cdot 0052891 \sin^2\varphi - 0 \cdot 0000059 \sin^2 2\varphi) \quad (3.21)$$

which corresponds to $f = 1/297 \cdot 1$, and Uotila's (1957):

$$\gamma_0 = 978 \cdot 0496(1 + 0 \cdot 0052934 \sin^2\varphi - 0 \cdot 0000059 \sin^2 2\varphi) \quad (3.22)$$

which gives $f = 1/297 \cdot 4$.

It should be emphasized that the International Formula is still adopted as the standard for gravity reductions.

None of the above formulae contains a longitude term, that term having been set equal to zero in the development of Clairaut's equation. It is quite probable that there are very widespread departures of gravity from the smoothed variation, and therefore departures of the geoid from the spheroid, which give the sea-level surface the general form of a tri-axial ellipsoid. Gravity formulae containing a longtitude term have been developed (Zhongolovich, 1952), and these correspond to a reference surface whose equatorial plane is elliptical, with a flattening of about 1/30,000. However, the present tendency is to maintain the spheroid as the reference surface and to treat any ellipticity of the equator with other geoidal warpings.

The Undulations of the Geoid

The effect of lateral variations in density on the equipotential surfaces of the gravitational field is indicated in Fig. 3.3. Over a region of mass excess, there is an additional potential ΔU, and the equipotential surface is warped outward. For a single mass anomaly in an otherwise uniform earth,

$$gN = \Delta U, \quad (3.23)$$

where N is the warping of the geoid (AB), and g is the mean value

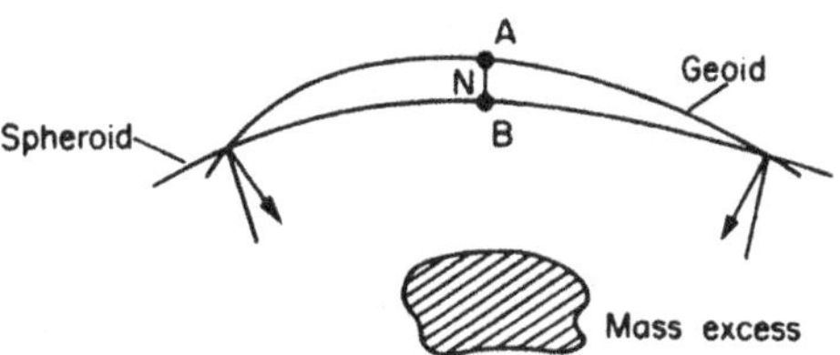

FIG. 3.3. Undulation of the geoid and deflection of the vertical caused by a mass anomaly.

of gravity between A and B. On either side of the region of mass excess, the vertical is deflected inward. A mass deficiency in the earth has the opposite effect.

The problem in the case of the actual earth is that there are a great number of mass anomalies contributing to the value of N at any point. In general, neither the masses themselves, nor their disturbing potential, are known, but the effect of the masses on g is. Mass anomalies, or lateral differences in density, cause g at sea level to differ from the value predicted by the smoothed formulae. A means of determining N from the anomalies in gravity remains to be found. As before, we neglect for the present the effect of mass above sea level, and assume that g is known at sea level over the earth.

The relation between N and the gravity anomalies, Δg, originally derived by Stokes (1849), has been obtained in a number of ways. All of these involve the expansion of both quantities in a series of spherical harmonics (Appendix 1), but the result is obtained most directly by a method by Pizetti (1911), which makes use of equation (49), Appendix 1.

At a point P on the geoid, at height N above the spheroid, let the gravity anomaly (observed $g - \gamma_0$ for that latitude) be Δg, and let the potential due to all disturbing masses be U. Δg arises partly from the attraction of the anomalous masses, and partly from the fact that the observer at P is at a distance N farther from the earth's centre than a corresponding point on the spheroid. It will be shown in the next chapter that the latter effect is given by

$$-\frac{2g}{r}N = -\frac{2U}{r}, \tag{3.24}$$

where r is the radius vector to P.

$$\therefore \Delta g = -\frac{\partial U}{\partial r} - \frac{2U}{r}. \tag{3.25}$$

Let
$$V_p = -r\Delta g = 2U + r\frac{\partial U}{\partial r} = \frac{1}{r}\frac{\partial(Ur^2)}{\partial r}. \tag{3.26}$$

So that
$$Ur^2 = \int rV_p dr + C. \tag{3.27}$$

V itself is a potential quantity, and by equation (49) of Appendix 1, V_p is given in terms of its values V_s over a sphere of radius a by

$$V_p = a\frac{(r^2 - a^2)}{4\pi}\int\frac{V_s}{d^3}d\sigma, \tag{3.28}$$

where $d^2 = a^2 + r^2 - 2ar\cos\psi$, ψ being the angle between the radius vector to P and that to the variable element of surface of the sphere. The integration is to be carried out over the complete sphere, with $d\sigma$ an element of solid angle.

$$\therefore Ur^2 = \frac{a}{4\pi}\int V_s d\sigma \int\frac{r^3 - ra^2}{d^3}dr + C. \tag{3.29}$$

Now
$$V_s = 2U + a(\partial U/\partial r)_a = -a\Delta g. \tag{3.30}$$

Substitution of (3.30) into (3.29), and integration with respect to r, leads to

$$U = \frac{a}{2\pi}\int \Delta g f(\psi)d\sigma, \tag{3.31}$$

where
$$f(\psi) = [\tfrac{1}{2}\operatorname{cosec}\tfrac{\psi}{2} - 1 - \cos\psi]$$
$$+ 3[1 - \cos\psi - 2\sin\tfrac{\psi}{2} - \cos\psi\log_e(\sin\tfrac{\psi}{2} + \sin^2\tfrac{\psi}{2})].$$

From which we obtain

$$N = \frac{a}{2\pi g}\int \Delta g f(\psi)d\sigma \tag{3.32}$$

which is Stokes' formula.

Here, a and g are to be thought of as mean values of the radius of the spheroid, and of gravity on it.

Stokes' formula gives the value of N at a point on the geoid in terms of the values of Δg over the entire earth. No assumption is required regarding the disturbing masses, except that they must be inside of the geoid. The function $f(\psi)$ is a weighting factor, which weights the anomalies according to their angular distance ψ from P. The behaviour of $f(\psi)$ as ψ varies from 0 to π is obviously of great importance, as it determines the degree to which distant mass anomalies influence the calculation of N at some point.

However, it is more convenient for actual calculations to work in terms of $\mathrm{d}\psi$ rather than $\mathrm{d}\sigma$. In this case

$$N_P = \frac{a}{2\pi g} \int_0^{2\pi} \mathrm{d}A \int_0^{\pi} \Delta g f(\psi) \sin\psi \, \mathrm{d}\psi, \qquad (3.33)$$

where A is the azimuth from P to the point to which Δg refers.

The weighting function is now $f(\psi)\sin\psi$, and this is plotted in Fig. 3.4. The relatively slow convergence to zero is a fact which

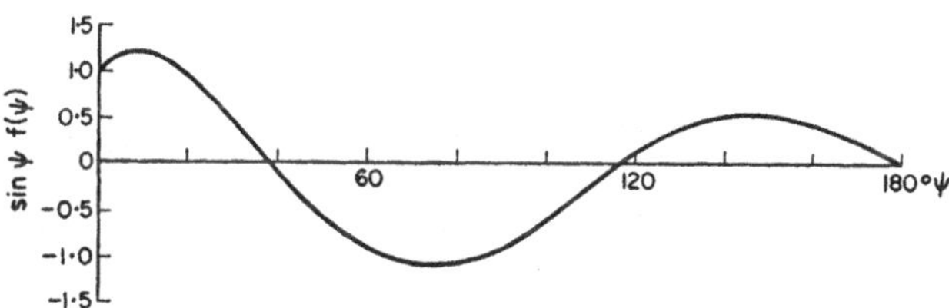

FIG. 3.4. The function $f(\psi) \sin \psi$.

has plagued geodesists from the time Stokes published his result. The calculation of N at any point is in fact dependent on a knowledge of Δg over the whole earth, and even today there are barely sufficient observations in some regions to permit an accurate calculation anywhere.

If there were a good distribution of points with known Δg, we could proceed as follows. Describe zones about P, with increments of angular distance $\Delta\psi$. Let

$$\Delta g_\psi = \frac{1}{2\pi}\int_0^{2\pi} \Delta g\, dA \qquad (3.34)$$

be the mean value of Δg on a circle of angular distance ψ. Then

$$N = \frac{a}{g}\int_0^\pi \Delta g_\psi f(\psi)\sin\psi\, d\psi. \qquad (3.35)$$

If $\overline{\Delta g_\psi}$ is the mean value of Δg_ψ in the interval between two circles, corresponding to ψ_1 and ψ_2,

$$N = \frac{a}{g}\sum \overline{\Delta g_\psi} f'(\psi), \qquad (3.36)$$

where

$$f'(\psi) = \int_{\psi_1}^{\psi_2} f(\psi)\sin\psi\, d\psi.$$

The function $f'(\psi)$ has been tabulated by Lambert and Darling (1936), so that the evaluation of (3.36) would present no difficulties if reliable mean values of Δg were available. We shall delay a discussion of the results that have been obtained with Stokes' formula until the problem of reducing continental observations of g to sea level is treated.

Expansion of the External Field in Spherical Harmonics

Stokes developed his expression for N in the form shown in equation (3.32) to avoid the labour of expanding the field into spherical harmonic terms. With an expansion for the gravity anomalies available, the corresponding harmonics in the expansion of N can easily be obtained. If n is the degree of a spherical harmonic in the expansion of the anomaly field the corresponding term in the expansion for N can be shown to be $a/g(n-1)$ times as large. There appear to be several advantages now to proceed by spherical harmonic expansion from the start. The computation is not impracticable when high-speed computers are used, conditions can be imposed for the vanishing of forbidden harmonics (such as the first), and account can be taken of the important information provided by observations on satellite orbits.

The usual procedure is to obtain average values for Δg over "squares" of $10°$ latitude by $10°$ longitude, and then fit to these a spherical harmonic expansion up to the desired order. By this means, Jeffreys (1943) and Zhongolovich (1956) obtained terms up to order 3. Kaula (1959) made use of all gravity data available up to 1958, and, in addition, the conditions imposed by satellite observations up to the same year, to obtain an expansion for Δg up to order 8.

Deflection of the Vertical

The deflections of the vertical at a point P, in north–south and east–west vertical planes, are equal to the slopes of the geoid, and therefore to the derivatives of N in these directions (Fig. 3.1). With the usual convention of signs, the deflections η and ξ can be written

$$\left. \begin{array}{l} \eta = -\dfrac{\partial N}{\partial x} \\[3mm] \xi = -\dfrac{\partial N}{\partial y} \end{array} \right\} \tag{3.37}$$

where x is toward the north, and y toward the east.

Operating on equation (3.33) gives the deflections in radians as

$$\left. \begin{array}{l} \eta = -\dfrac{1}{2\pi g} \displaystyle\int\int \Delta g \,\dfrac{\partial f(\psi)}{\partial \psi}\sin\psi\cos A \,d\psi \,dA \\[5mm] \xi = -\dfrac{1}{2\pi g} \displaystyle\int\int \Delta g \,\dfrac{\partial f(\psi)}{\partial \psi}\sin\psi\sin A \,d\psi \,dA \end{array} \right\}. \tag{3.38}$$

The quantity $[\partial f(\psi)/\partial \psi] \sin \psi$ has been tabulated by Sollins (1947), and the numerical evaluation of equation (3.38) can be carried out in a manner similar to that described for N. The contribution of distant zones decreases more rapidly in the case of the deflections than it does for N itself. A deficiency of measurements over certain regions of the earth is therefore less serious in this calculation, even though the absolute value of N can only be obtained by the complete evaluation of equation (3.33). The calculation of the deflection of the vertical from gravity observations is particularly valuable at base stations of geodetic networks. Rice (1951) has described the calculation for a number of stations in the United States, including Meade's Ranch, Kansas, the base for the North American geodetic system.

The Contribution of Satellite Observations

Just as the description of the motions of planets by Kepler led Newton to formulate the universal law of gravitation, so the observations that have been made on artificial satellites in the years since 1957 have been extremely valuable in determining the earth's gravitational field. If the earth were a uniform sphere, with a spherically symmetrical gravity field, satellites would describe elliptical orbits about it, according to Kepler's laws. Because of the departure of the field from this condition, the orbits are perturbed. Observations on the actual orbits, corrected for any effect due to air resistance, can therefore be used to determine spherical harmonic terms in the field (Kaula, 1962). The method of approach is identical to that which has been used in celestial mechanics for many years, and an outline of it is given in Appendix 3.

It is evident that the greatest effect on an orbit will be that produced by the lowest harmonics in the gravitational field. Satellite observations therefore led first to a redetermination of the second harmonic, related, as we have seen, to the flattening of the spheroid. Observations on Sputnik 2 were analysed by Merson and King-Hele (1958), and those on Vanguard I by O'Keefe, Eckels and Squires (1959). Merson and King-Hele noted that the rate of rotation of the orbital plane (designated by

$\dot{\Omega}$ in Appendix 3) of Sputnik 2 was less than that predicted for a flattening of $1/297 \cdot 1$. The two rates could be made to agree if the flattening were taken as $1/298 \cdot 24$. Satellite observations thus indicate that the values of f previously obtained, including those based on gravity observations, are too large. If a spheroid of flattening $1/297$ is retained as a reference figure, the field determined by the satellites would require the presence of extensive geoidal bulges, of amplitude approximately 80 m, over the polar regions of the earth. This has led to the description of the earth's sea-level figure as "pear-shaped". It should be kept in mind, however, that these polar bulges of 80 m are really rather minute when compared to the main flattening of 20 km.

CHAPTER 4

The Reduction of Gravity Observations

Variation of Gravity with Height

Gravity on the continents is only rarely measured at sea level, and both geodetic and geophysical interpretations of the measurements require that a correction be made for the height of the station. The variation in g due to change in distance from the earth's centre is obtained immediately by differentiation. If the earth is assumed to be spherical, and of mass M, the value of g at a point distant r from the centre is

$$g = \frac{GM}{r^2}, \tag{4.1}$$

$$\therefore \frac{\partial g}{\partial r} = -2\frac{GM}{r^3} = -\frac{2g}{r}. \tag{4.2}$$

At sea level, equation (4.2) gives the vertical gradient of gravity as $-0\cdot3086$ mgal/m, or $-0\cdot09406$ mgal/ft. For most purposes, this value may be used anywhere on earth. A more complete evaluation, taking into account the spheroidal shape, gives

$$\frac{\partial g}{\partial r} = -0\cdot30855 - 0\cdot00022\cos2\varphi + 0\cdot000144h \tag{4.3}$$

in mgal/m, where φ is the geocentric latitude and h is the height in metres. Equations (4.2) and (4.3) give the rate at which g decreases with increasing distance from the earth's centre, or height, if no additional mass is interposed between the observer and the earth. The decrease is that which would be measured by an observer rising in a balloon through the air, and the gradient is therefore known as the free-air gradient. If g is measured on the land surface at different heights, at the same latitude, the variation with height

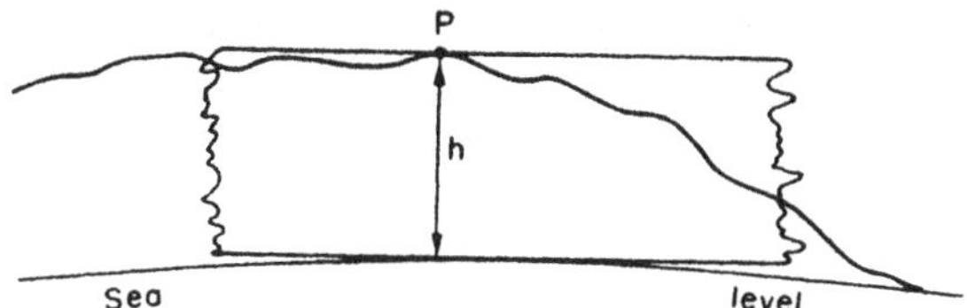

FIG. 4.1. The Bouguer approximation for attraction of mass above sea level.

will not be found to be that given by the free-air term, because of additional mass beneath the higher stations. Bouguer (1749) realized this during the course of his measurements in the Andes, and suggested that the additional attraction due to material above sea level be approximated by treating this material as an infinite horizontal slab (Fig. 4.1), of thickness equal to the height of the station. The attraction of such a slab, which we shall derive in a subsequent chapter, is

$$\Delta g = 2\pi G \varrho h \qquad (4.4)$$

in gal, where ϱ is the density of the material and h is in centimetres. The term Δg here represents a positive contribution to g at the higher station, and the Bouguer effect is therefore of opposite sign to the free-air decrease. For a typical density of crustal material, $2 \cdot 67$ g/cm^3, the Bouguer term is $0 \cdot 1118$ mgal/m. This is less than the free-air gradient, consequently g measured on the land surface does decrease with increasing height, in this case at the net rate of $0 \cdot 1968$ mgal/m. The validity of the Bouguer approximation in the case of stations located in areas of rugged terrain may be questioned, but it is in fact most convenient to use his formula as it stands, and to add a correction, usually small, for the departure of the terrain from a plane. The calculation of this correction is described later.

Variation of Gravity within the Earth

The actual variation of g inside the body of the earth does not enter into the reduction of gravity observations, but it is convenient to discuss it here. At a point inside a uniform sphere, at distance r from the centre, the theory of potential (Appendix 1) shows that there is no resultant attraction from that portion of the sphere which lies outside the surface of radius r. The attraction arises from a spherical mass of radius r, and decreases linearly toward the centre. In the actual earth, the density is by no means uniform, but we may approximate the variation by a series of concentric spherical shells. (The actual shape of surfaces of equal density is mentioned in Appendix 2.) As the attraction of each shell is equivalent to that of a point mass at the centre, the attraction at any radius r is

$$g_r = \frac{Gm}{r^2},\tag{4.5}$$

where m is the total mass within the sphere of radius r. Since we are interested here only in the variation of average g with depth, the effect of the earth's rotation is neglected.

The variation of density within the earth has been investigated in detail by Bullen (1953). If the interior is assumed to be in a hydrostatic state, the pressure gradient at radius r is

$$\frac{dp}{dr} = -g_r \, \varrho_r,\tag{4.6}$$

where ϱ_r is the density at that level. Now, for adiabatic conditions

$$\frac{dp}{d\varrho} = \frac{K}{\varrho},\tag{4.7}$$

where K is the bulk modulus. It is convenient that the ratio K/ϱ can be determined in terms of the seismic wave velocities, which are well known as functions of depth. In particular,

$$\frac{K}{\varrho} = V_p^2 - \tfrac{4}{3}\, v_s^2,\tag{4.8}$$

where V_p and v_s are the velocities of longitudinal and transverse waves respectively. We have, therefore,

$$\left(\frac{d\varrho}{dr}\right)_r = \frac{d\varrho}{dp}\cdot\frac{dp}{dr} = \frac{Gm\varrho_r}{r^2(V_p^2 - v_s^2)}. \tag{4.9}$$

Equation (4.9) is evaluated numerically, starting at the top of the mantle, where the quantity m is the mass of the earth less the mass of the crust. The density at the top of the mantle must be assumed, after which the gradient is computed. This allows the density at a depth of, say, 100 km to be estimated, and the process is repeated. The final determination of the density variation is controlled by the known mass of the earth, and the moment of inertia about the axis of rotation. This is important, since new assumptions must be made as to the density below any discontinuities. The quantity m is determined for each level at which equation (4.9) is evaluated, so that g_r can be calculated from equation (4.5). Figure 4.2 shows the variation in gravity with depth. The remarkable feature is the constancy of g in the outer part of the earth, in contrast to the linear decrease which would take

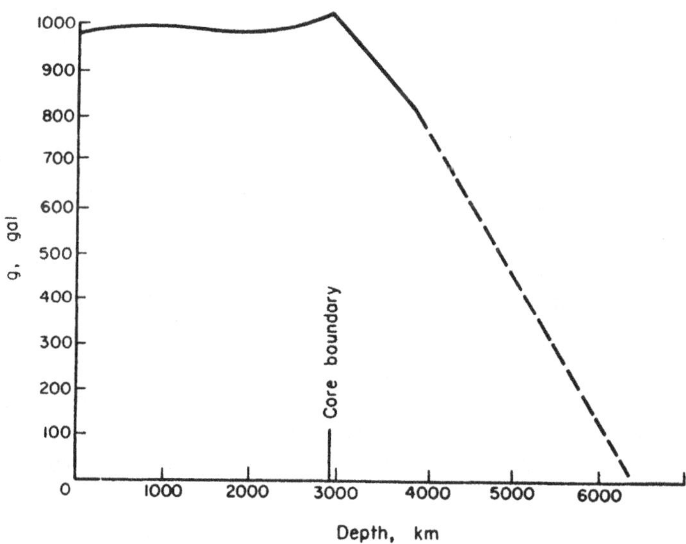

FIG. 4.2. Variation of gravity within the earth.

place in a uniform sphere. The concentration of mass in the core produces an attraction which increases with depth, compensating for the vanishing attraction of the outer layers.

Gravity Anomalies

A gravity anomaly is the difference between the measured value of g at some point, and a theoretical value, usually that predicted by the international formula, for the same point. If g_0 is the measured value on the land surface, at height h, it must first be corrected to sea level before it can be compared to γ_0, the theoretical value, for the same latitude.†

Suppose first that g_0 is corrected to sea level by use of the free-air term only, although this may not seem appropriate at the moment. The free-air anomaly $\varDelta g_F$ is then

$$\varDelta g_F = (g_0 + 0 \cdot 3086h \times 10^{-3}) - \gamma_0, \qquad (4.10)$$

where h is in metres, and $\varDelta g$, g_0 and γ_0 are in gal.

On the other hand, if g_0 is reduced to sea level by means of Bouguer's correction, the Bouguer anomaly $\varDelta g_B$ is obtained, where

$$\varDelta g_B = [g_0 + (0 \cdot 3086 \times 10^{-3} - 200\pi G\varrho)h] - \gamma_0. \qquad (4.11)$$

It will be noticed that the calculation of the Bouguer anomaly requires an assumption as to the mean density, ϱ, of material between sea level and the station height.

Variation of Free-air and Bouguer Anomalies over the Earth

It might be expected that, since in the calculation of free-air anomalies no account is taken of the attraction of material above sea level, these anomalies would tend toward large positive values at higher stations. That this is indeed the case is shown by the following table from Bowie (1917). It is based on the average anomalies at a number of stations in the United States.

† Tabulated values of γ_0, for every 10′ of latitude, were computed by Lambert and Darling (1931), and reprinted in Nettleton (1940). Tables are also given in Jung (1952*b*).

TABLE 4.1

	Mean anomalies with regard to sign (mgal)	
	Bouger	Free-air
Coast stations	+ 17	+ 17
Inland stations, not in mountainous areas	− 28	+ 9
Stations in mountainous areas, below general level	− 107	− 8
Stations in mountainous areas, above general level	− 110	+ 58

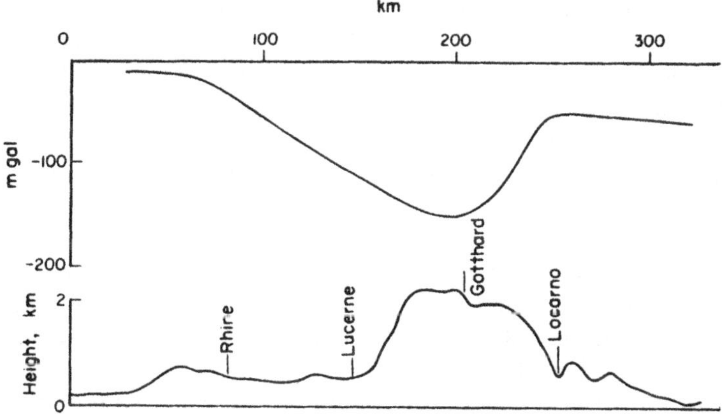

FIG. 4.3.   Profile of Bouguer anomaly across the Alps.

The free-air anomalies are positive for the higher stations, but even more striking is the tendency of the Bouguer anomalies to go to large negative values. Gravity measurements have been made in most of the mountainous regions of the world, and many examples (Fig. 4.3) could be given of this almost universal tendency. Now there is nothing illogical in the inclusion of the Bouguer term for the attraction of material above sea level, and the large negative values can only mean that there is a deficiency in mass below sea level, with the deficiency increasing as the land surface rises. This is an exemplification of the important principle of the compensation of the earth's surface features by sub-surface mass distributions. As this principle was suggested before detailed measurements of the intensity of gravity were available, we shall look briefly at the historical development of the ideas.

Isostasy: Compensation of Surface Features

Many writers, as early as the seventeenth century, suggested that mountains stand up because of lighter material beneath them. Attempts to determine the gravitational constant G during the eighteenth century by measuring the attraction of mountains were plagued by the "concealed masses". However, definitive ideas date from the analysis of deflections of the vertical measured during the survey of India. Stations in the vicinity of the Himalayas did indeed exhibit deflections toward the mountains, as judged by the discrepancies between geodetic and astronomic positions. However, Pratt (1855) computed the deflections to be expected from the mass of the mountains, and showed that the observed effect was only about one-third of that expected. Airy (1855) interpreted these comparisons as proof of compensation of the mountain masses, and suggested the presence of a fairly thin crust, resting on a denser sub-stratum. Beneath elevated regions the crust projects downward as roots, so that the total mass per unit area down to some level beneath the deepest roots is everywhere the same (Fig. 4.4). Pratt (1859) agreed with the idea of

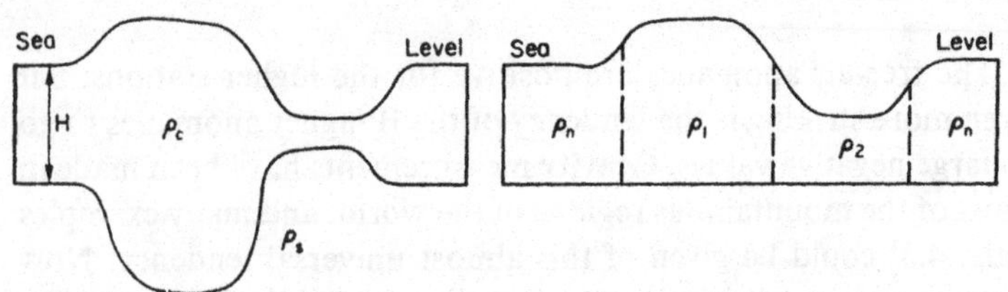

FIG. 4.4.   Assumptions of Airy (left) and Pratt isostasy.

compensation but did not accept Airy's explanation of the mechanism. He proposed a crust, extending to a uniform depth below sea level, in which the density varied inversely as the height of the topography (Fig. 4.4). Pratt had in mind the idea that mountains were formed by the vertical expansion of crustal columns, with no change in mass. Years later, Dutton (1889) introduced the word "isostasy" to describe the condition of compensation, and the presence of a hydrostatic state beneath a certain depth within the earth. About the same time investigations into the "theory of isostasy" began to be made with measurements of g as well as with plumb-line deflections. For many years the emphasis was on the testing of one or other hypothesis, Airy's or Pratt's, and on the attempt to find the best parameters, such as density and crustal thickness, in each case. The general approach (Hayford and Bowie, 1912; Heiskanen, 1938) was to compute the effect on gravity of the compensating masses, inferred from the visible topography for either theory, and to remove this from the Bouguer anomaly. The attempt was then made to reduce the resulting "isostatic anomalies" by a suitable adjustment of parameters.

Calculation of Isostatic Anomalies

In order to calculate the effect on gravity at a station of the masses compensating the topography, Hayford and Bowie (1912) divided the region surrounding the station into compartments, each having the form of a cylindrical segment. The attraction of mass in this form, for a point on the axis of the cylinder, can readily be obtained. We will leave the detailed calculation to Chapter 5, and write down the result. With the notation of Fig. 4.5, the contribution to g, at the point P, due to the prism of mass shown, is

$$\Delta g = \varphi G \Delta\varrho \left[(r_1^2 + l_2^2)^{1/2} - (r_1^2 + l_1^2)^{1/2} - (r_2^2 + l_2^2)^{1/2} + (r_2^2 + l_1^2)^{1/2} \right],$$

$$(4.12)$$

where, as usual, Δg is in gal, all distances are in cm, φ is in radians, and $\Delta\varrho$ is the anomalous density.

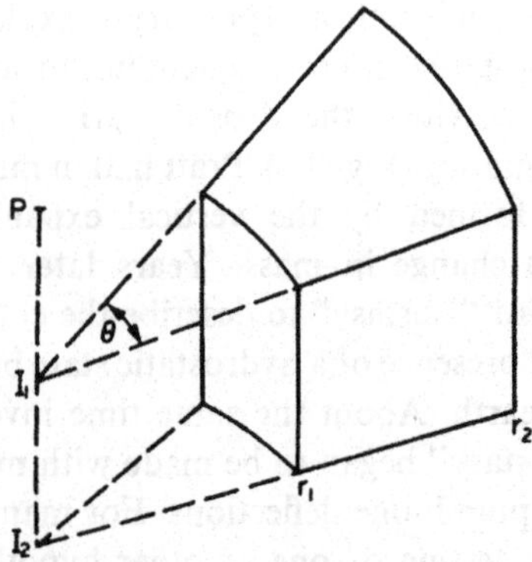

FIG. 4.5.　A segment of the crust as defined for isostatic calculations.

Now suppose the mean elevation, above sea level, of the surface of the segment is h. Then on Pratt's hypothesis, the density of the material in the prism will not be ϱ_n, the normal density of the earth's crust, but will be

$$\varrho = \varrho_n \cdot \frac{H}{H+h}, \qquad (4.13)$$

where H is the constant depth to which the Pratt crust extends below sea level.

The anomalous density is then

$$\Delta\varrho = \varrho - \varrho_n \qquad (4.14)$$

$$= \varrho_n\left(\frac{H}{H+h}-1\right) = \frac{-h\varrho_n}{H}$$

very nearly.

If this value of $\Delta\varrho$ is inserted into equation (4.12) the effect on g at the point P, of the compensation of the topography in this one segment is obtained. Some care is required if the surface height of the segment, h, is greater than the height of the point P, for equation (4.12) as it stands is valid only if all the mass is below P. The total effect of compensation of all topographic

TABLE 4.2. HAYFORD ZONES

Zone	Outer radius(m)	Number of compartments	Zone	Outer radius	Number of compartments
A	2	1	18	1°41′13″	1
B	68	4	17	1°54′52″	1
C	230	4	16	2°11′53″	1
D	590	6	15	2°33′46″	1
E	1 280	8	14	3°03′05″	1
F	2 290	10	13	4°19′13″	16
G	3 520	12	12	5°46′34″	10
H	5 240	16	11	7°51′30″	8
I	8 440	20	10	10°44′	6
J	12 400	16	9	14°09′	4
K	18 800	20	8	20°41′	4
L	28 800	24	7	26°41′	2
M	58 800	14	6	35°58′	18
N	99 000	16	5	51°04′	16
O	166 700	28	4	72°13′	12
			3	105°48′	10
			2	150°56′	6
			1	180°00′	1

features over the earth is obtained by adding the contribution of all segments, which in Hayford and Bowie's analysis extended to the antipodes of the station. (Equation (4.12) must be corrected for the curvature of the earth in the case of distant segments.) Once the size of the segments, and the values of H and ϱ_r, have been fixed, the contribution of each segment may be tabulated in terms of h. Table 4.2 gives the radii of the zones, and the number of compartments in each, for the system used by Hayford and Bowie. These are drawn on transparent templates at scales corresponding to topographic maps available. The greatest labour involved in making the isostatic reduction is then the estimation of the heights of the compartments. For compartments lying over the oceans, h is negative, but a different density factor is used because of the mass of sea water. Hayford and Bowie took

$$\Delta\varrho = -\frac{h}{H}\,2\cdot67$$

for land compartments, and

$$\Delta\varrho = -\frac{h}{H}\,1\cdot64$$

for sea compartments.

The sum of the effects of all compartments is added to the Bouguer anomaly to give the Pratt (or Hayford) isostatic anomaly. Different anomalies are possible, depending chiefly on the value chosen for H. Hayford and Bowie (1912) found that for stations in the United States a value of H of $113\cdot7$ km gave the smallest sum of squares of anomalies, and suggested that this was the best value of the "depth of compensation" on Pratt's hypothesis.

On the Airy hypothesis, compensation is achieved by variations in the thickness of a crust of uniform density. The effect on g of the compensating masses can be computed in this case also, although this was not done by Hayford and Bowie. Heiskanen (1924) first showed that the same system of compartments used by these workers could be used to test the Airy theory. We let H in this case be the thickness of the Airy crust for a segment at sea level, ϱ_c the constant density of the crust, and ϱ_s the density of the sub-stratum. Then, for a segment of height h above sea level, the downward projection of the root, h' (Fig. 4.4), is obtained from the relation

$$\varrho_c H + \varrho_s h' = \varrho_c(h + H + h')$$

$$h' = h\cdot\frac{\varrho_c}{\varrho_s - \varrho_c}. \qquad (4.15)$$

The effect of the compensating root for any segment can again be obtained from equation (4.12), in which l_1 will be approximately H, and l_2, $H+h'$ (there will be a small correction for the height of the station itself). Once again, if H, ϱ_c and ϱ_s are adopted, the contribution of each segment can be tabulated in terms of h, the mean elevation of that segment. This was done by Heiskanen (1938). The sum of the contributions of all segments, added to the Bouguer anomaly, gives the Airy or Heiskanen anomaly, of which a number may be computed, depending chiefly on the value of H assumed. Heiskanen's investigations have shown that values of H between 20 and 40 km give the smallest isostatic anomalies.

Terrain effects. The methods of reduction outlined above for Bouguer and isostatic anomalies have assumed that the land surface surrounding the station is a horizontal plane, of height equal to the station elevation. In areas of rugged terrain this is not the

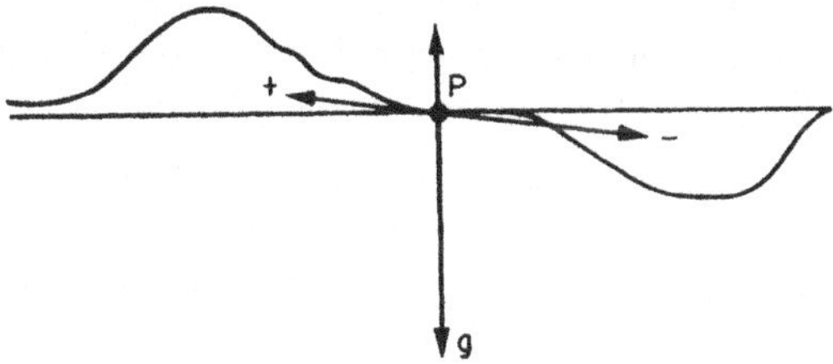

FIG. 4.6.   The effect of local terrain on gravity.

case, and a correction must be made for hills and valleys in the vicinity of the station. Figure 4.6 illustrates the situation. Since both mass above and deficiency of mass below a station tend to reduce the value of g, the correction to be applied is always positive. The attraction of the topography is most conveniently estimated by dividing the region around the station into segments, as was done for the estimate of compensation. The terrain effect for any segment is then a function of the difference, whether positive or negative, between the station height and the mean height of the segment. It is possible to underestimate the correction if too coarse a system of compartments is used. For example, the mean elevation of a compartment may turn out to be the same as the station height, in which case there would be no terrain effect indicated for it. But the compartment may contain hills

C

and valleys whose effect is appreciable. For local gravity surveys, where the Bouguer anomaly only is required, but the terrain correction is desired with high precision, the Hammer (1939) system of small compartments will generally be found more convenient than the Hayford zones. In either case the estimation of mean elevations from topographic maps demands some skill in visualizing the features of the topography. Only if the topography within a compartment is in the form of a uniform slope is the desired mean simply the numerical average of greatest and least height.

The effect of terrain within 167 km of a station (that is, within the outer radius of Hayford zone O) may exceed 100 mgal for mountain stations on peaks or in narrow valleys. It is normally less than 50 mgal for stations even in mountainous areas, and rarely exceeds 1 mgal for stations outside of these areas.

Reductions for sea stations. In the past, most gravity measurements at sea were made with the Vening Meinesz pendulums in submerged submarines. For these stations, the observed gravity must be corrected upward to sea level, by the subtraction of $0 \cdot 3086\ d$, where d is the depth of the submarine in metres, and the addition of a term equal to twice the attraction of the layer of water above the submarine, that is, $4\pi G(1 \cdot 027)d \times 10^2$. The difference between the observed value, thus corrected, and γ_0 gives the free-air anomaly for sea stations. If, in addition, the ocean beneath the station is replaced by normal crustal rock, through the addition of a term $4\pi G(2 \cdot 67 - 1 \cdot 027)t \times 10^2$, where t is the depth of the ocean in metres, a quantity equivalent to the Bouguer anomaly for land stations is obtained. Because the mass deficiency of the oceans is compensated by denser material beneath them, the Bouguer anomalies at sea stations are characteristically strongly positive.

Choice of Anomaly to be used for Various Purposes

We have introduced the free-air, Bouguer, and various isostatic anomalies. It may well be asked at this point what use can be made of these reductions. The applications with which we are concerned include: the determination of the flattening of the spheroid by Clairaut's theorem, or its extensions; the calculation of undulations of the geoid by Stokes' formula; the test of theories of isostasy, and the investigation of mass anomalies in the earth for various purposes. In the derivation of Clairaut's theorem, it was assumed that there was no mass external to the spheroid, and that values of g were available on that surface. To use measurements made on the land surface, we must neglect the material above sea level, and correct g to sea level by the free air reduction. A further free-air correction should then be made for the departure of the geoid from the spheroid below the point of observation. Tables for estimating this small "indirect effect" have been published (Lambert and Darling, 1936). Neglect of the attraction of material above sea level in this case need cause no concern here. The continents have in effect been replaced by additional mass located below sea level, but this mass will not affect the determination of the main flattening.

For the test of particular theories of isostatic compensation, the corresponding isostatic anomalies are convenient. The combination of parameters which yields the smallest mean anomaly over a region is generally taken as indicating the best model for the particular theory under study, although the obtaining of a small mean anomaly must not be taken as proof that the model is correct. For investigation of mass differences in the crust, particularly those of limited extent, the Bouguer anomaly is the most appropriate to use. It eliminates, insofar as is possible, differences in g between neighbouring stations which are due to differences in height.

There remains the reduction to be used in the application of Stokes' formula to the determination of geoidal undulations. In this case there has been much difference of opinion between geodesists. The derivation of Stokes' equation assumed that there was no mass external to the geoid, and that g on the geoid was known. It is agreed that the mass external to the geoid should be removed in the gravity reduction, but as this may change the equipotential surface, a final step may be necessary to revert to the "natural geoid". However, Helmert (1884) pointed out that a simple condensation of the mass above sea level, to a layer just inside the geoid, would in fact change that surface by a negligible amount. The reduction to be used in this case is then just the free-air one, as the attraction of the actual material above sea level is virtually the same as the attraction of the layer at a point on the geoid. The layer or coating at any point would have a surface density of ϱh g/cm^2, where ϱ is the density of the topography and h is the height of the station above sea level. At a point on the geoid, below the station, the attraction of this coating would be $2\pi G\varrho h$, which is identical (apart from any terrain effect) to the attraction of the actual topography at the station. The most direct approach in the application of Stokes' equation would thus appear to be the use of free-air anomalies. They have the disadvantage that their values are strongly correlated with the actual station heights, and when a mean anomaly for a region is determined for use in the calculation, it must be that corresponding to the mean height. In mountainous areas, most stations are in valleys, and the free-air anomaly corresponding to the mean height may be difficult to obtain.

Other geodesists (Heiskanen, 1957) have argued strongly for the use of isostatic anomalies in geoidal determinations. It is certainly easier to estimate average values for these over an area, as they do not vary markedly with station height. On the other hand, their use in Stokes' formula is equivalent to the assumption that only uncompensated masses contribute to the undulations of the geoid, which may not be precisely the case.

It should be remembered that the ultimate aim of this branch of geodesy is to describe the true form in space of the earth's outer surface. A figure other than the geoid as defined above would be equally satisfactory as a reference, provided the heights of the land surface above it could be determined. In this connection, Molodensky, Eremeev and Yurkina (1960) have argued strongly for the use of a "quasi-geoid", a figure whose warpings are determined by integration of specially-defined gravity anomalies over the actual surface of the earth. These authors describe in detail the deter-

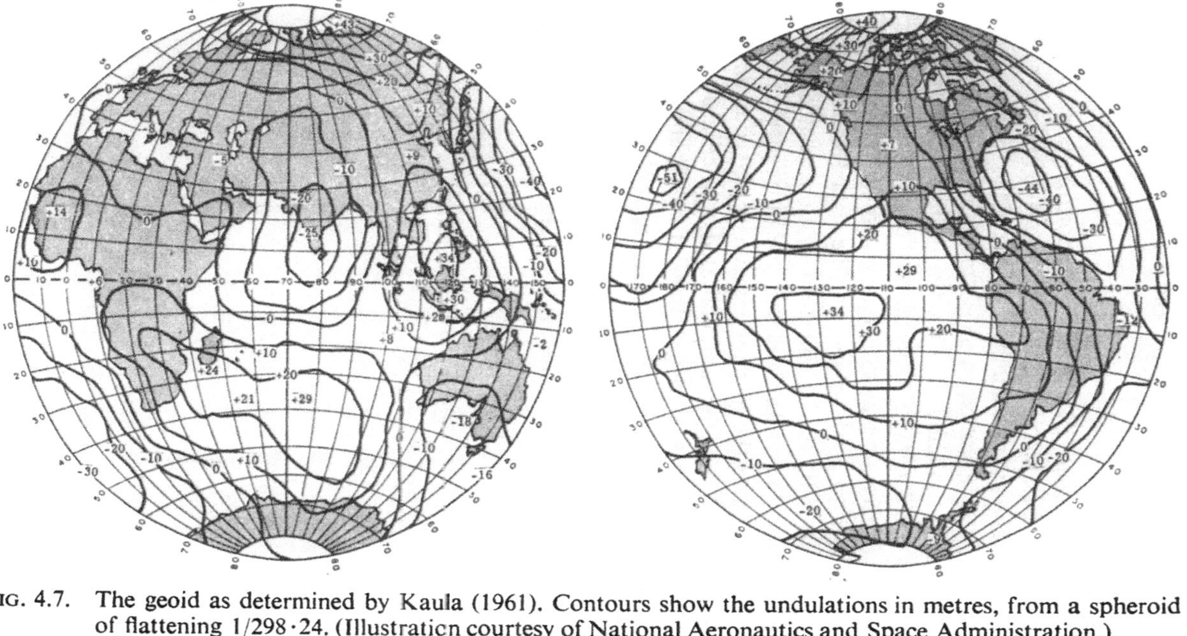

FIG. 4.7. The geoid as determined by Kaula (1961). Contours show the undulations in metres, from a spheroid of flattening 1/298·24. (Illustration courtesy of National Aeronautics and Space Administration.)

mination of both the quasi-geoid and the height of the land surface above it.

Examples of geoids. Recent determinations of the geoid, by rather different methods, include the Columbus geoid and one computed by Kaula. The first, computed at the Ohio State University (Heiskanen, 1957), was based on isostatic anomalies. Calculations were completed initially only for the northern hemisphere. The other is a geoid computed by means of a spherical harmonic expansion of free-air gravity anomalies to order 8, with control provided by astrogeodetic and satellite data (Kaula, 1961). As Kaula points out, the satellite observations available to him provided important information on the polar regions. It will be noted, Fig. 4.7, that the geoid in this case is referred to an ellipsoid of flattening 1/298·24, the value indicated by satellite observations. Undulations of the geoid of the order of 30–40, and, in one case, 50 m, are observed. The location, and amplitude, of the main undulations in the northern hemisphere is in fair agreement between the Columbus and Kaula geoids. These undulations have great significance for studies of the earth's interior, which will be discussed later.

CHAPTER 5

The Interpretation of Gravity Anomalies

General Principles

Anomalies in gravity, calculated by the methods discussed previously, are widely used to provide information on structures located beneath the earth's surface. Different rock types which occur in, and beneath, the earth's crust have different densities, so that mass is by no means uniformly distributed in the outer part of the earth. Over areas of mass excess, observed g is greater than normal, and there is a tendency toward positive anomalies. The aim of the geophysicist is to deduce from the pattern of anomalies the location and form of the structure which produces the disturbance in gravity. Normally, if the mass distribution is of primary interest, it will be the Bouguer anomalies which are considered, and these will be plotted in the form of profiles or contour maps.

There are two characteristics of the gravitational field which make a unique interpretation impossible. The first is that the measured value of g, and therefore the reduced anomaly, at any station reflects the superimposed influence of many mass distributions. The attraction of relatively local features is often seen only as a minor distortion of the pattern due to some major structure. Interpretation can only proceed after the contributions of different bodies are isolated. This effect is always present, but it becomes most serious in the case of geophysical prospecting, where extremely local structures are of interest. We shall therefore postpone the detailed consideration of methods of isolation to Chapter 8.

The second difficulty is that gravity, as a potential field, shares the fundamental ambiguity of inverse boundary value problems common to all potential fields (Appendix 1). For a given distribution of anomalies on (or above) the earth's surface, an infinite number of mass distributions can be found which would produce them. At first, the interpretation problem appears hopeless. However, geological reasonableness will often rule out whole classes of solutions, and other information, such as the probable density or depth of the source of the field, may lead one to the most likely mass distribution.

For many years, the accepted method of interpretation was to assume various simple shapes for the source of an anomaly, compute their effects at the surface, and modify them until a fit with the observed field was obtained. The achieving of a fit indicated only that the selected model was a possible solution. This cut-and-try process of interpretation is often called the indirect method of interpretation. In spite of its lack of elegance, it has many advantages. The computed effects of bodies of simple shape are readily available, and a quick comparison with the observed anomaly is possible. In the case of preliminary surveys, with observations of limited number or uncertain accuracy, this procedure may be all that is justified. On the other hand, if in a certain area there is complete coverage, with stations of high accuracy, it may be desirable to employ a more direct approach. Methods have been developed in recent years to extract information on the mass distribution by mathematical operations on the observed field. These methods, which generally require the use of high-speed computers, cannot reduce the fundamental ambiguity mentioned above. Some parameters of the unknown structure must be assumed at the start. Under the assumptions, they attempt to extract the maximum information from the field.

Indirect Methods

For the indirect approach, the interpreter must have available a selection of forms whose attraction can be computed. The anomalies due to some of the more useful of these are given on the following pages.

General Mass Distribution

The attraction at P due to an element of mass dm (Fig. 5.1) is Gdm/r^2, along PQ. The contribution to the gravity anomaly at P

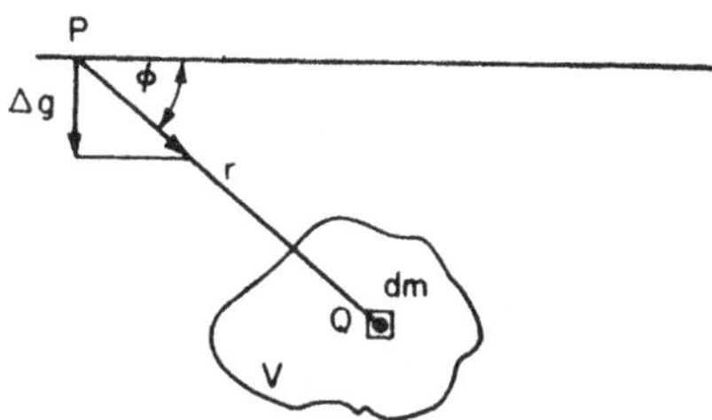

FIG. 5.1.   Calculation of gravity anomaly for a general mass distribution.

is the vertical component of this, or $(Gdm/r^2)\sin\varphi$. Here, dm is to be considered the excess mass contained in a volume element dv at Q, as compared to a volume element outside the body. If $\varDelta\varrho$ is the difference in density between the body and its surroundings, d$m = \varDelta\varrho \cdot dv$. The gravity anomaly due to the body is then

$$\varDelta g = G\varDelta\varrho \int_v \frac{\sin\varphi \, dv}{r^2}. \tag{5.1}$$

Sphere (Fig. 5.2)

The integration contained in equation (5.1) is not required for the sphere, as potential theory (Appendix 1) shows the external effect to be identical to that of a particle of equal mass at the centre.

$$\therefore \varDelta g = \tfrac{4}{3}\pi R^3 G\varDelta\varrho \cdot \frac{z}{r^3}$$

$$= \tfrac{4}{3}\pi R^3 G\varDelta\varrho \, \frac{z}{(x^2 + z^2)^{3/2}}. \tag{5.2}$$

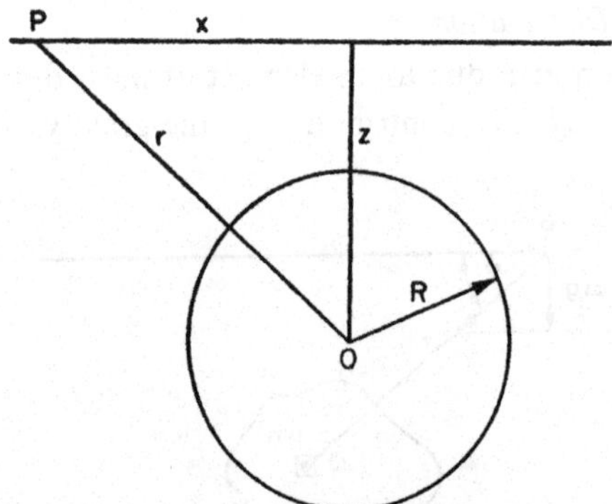

FIG. 5.2.　The sphere.

Vertical Cylinder, Point on the Axis (Fig. 5.3)

Consider first a cylinder extending to infinite depth. It is convenient to take cylindrical coordinates, with P as the origin. The element of volume is $rd\varphi drdz$, and the gravity anomaly is

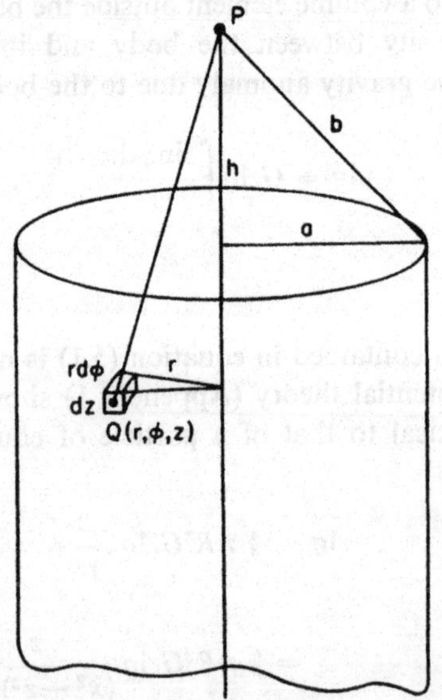

FIG. 5.3.　Vertical cylinder of infinite depth extent.

$$\Delta g = G\Delta\varrho \int\limits_{\phi=0}^{2\pi} \int\limits_{r=0}^{a} \int\limits_{z=0}^{\infty} \frac{zr\,d\varphi\,dr\,dz}{(r^2+z^2)^{3/2}}. \tag{5.3}$$

Integration gives, without difficulty,

$$\Delta g = 2\pi G\Delta\varrho\,[(h^2+a^2)^{1/2}-h]$$

$$= 2\pi G\Delta\varrho(b-h). \tag{5.4}$$

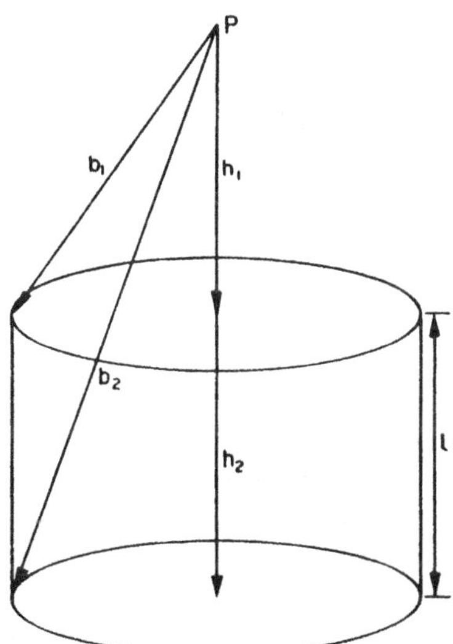

FIG. 5.4. Vertical cylinder of finite length.

For a cylinder of finite depth extent, the anomaly may be obtained from the difference of two infinite cylinders (Fig. 5.4):

$$\Delta g = 2\pi G\Delta\varrho\,[b_1-h_1-(b_2-h_2)]$$

$$= 2\pi G\Delta\varrho\,[l-(b_2-b_1)]. \tag{5.5}$$

If the radius of the cylinder becomes very large compared to h_1 and h_2, $b_2 - b_1 \to 0$, and

$$\Delta g = 2\pi G \Delta\varrho \,.\, l. \qquad (5.6)$$

This is the Bouguer formula for the attraction of a slab, which has already been mentioned.

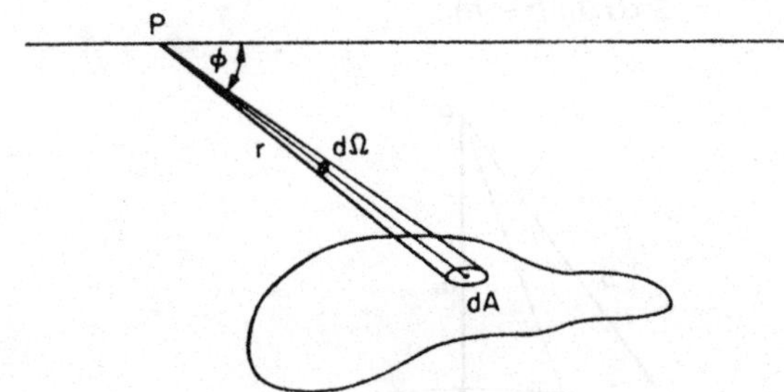

FIG. 5.5.   Horizontal sheet of mass.

Horizontal Sheet of Mass (Fig. 5.5)

Many structures are of sufficiently small vertical extent relative to their depth that their excess mass can be condensed on to a horizontal plane. If the actual density excess is $\Delta\varrho$, and the vertical thickness h, the equivalent excess surface density $\Delta\sigma$ is $h \,.\, \Delta\varrho$ g/cm². The contribution to Δg of the surface element dA is

$$\frac{G\Delta\sigma \sin\varphi \, dA}{r^2} = G\Delta\sigma \, d\Omega, \qquad (5.7)$$

where $d\Omega$ is the element of solid angle subtended at P by dA.

$$\therefore \Delta g = G\Delta\sigma\Omega, \qquad (5.8)$$

where Ω is the solid angle subtended at P by the sheet. Solid angles can be calculated for various outline shapes, and for circles, in particular, tables are available (Nettleton, 1942). These allow the gravity anomaly to be obtained very easily.

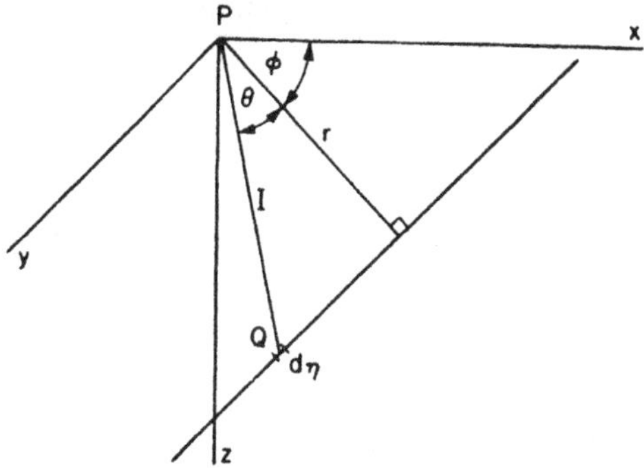

FIG. 5.6. Infinite horizontal line of mass.

Infinite Horizontal Line of Mass (Fig. 5.6)

The attraction of structures which are elongated in one direction (the geological strike), and of uniform cross-section, can be obtained from the infinite horizontal line. Let a mass m per unit length be distributed along the η-axis, which is parallel to y. At $P(0, 0, 0)$, the contribution to Δg due to the element $\mathrm{d}\eta$ at Q is

$$\frac{Gm}{l^2}\,\mathrm{d}\eta\,.\,\cos\theta\sin\varphi.$$

$$\therefore \Delta g = Gm\sin\varphi\int_{-\infty}^{\infty}\frac{\cos\theta}{l^2}\,\mathrm{d}\eta$$

$$= \frac{2Gm\sin\varphi}{r}. \tag{5.9}$$

For structures of this type, all of the operations may be confined to the x–z plane, and the bodies are said to be two-dimensional.

Horizontal cylinder (Fig. 5.7)

A horizontal cylinder of infinite length can be shown to be equivalent to a linear concentration of mass along the axis. The excess mass per unit length $= \pi R^2 \Delta \varrho$.

$$\therefore \; \Delta g = \frac{2\pi GR^2 \Delta\varrho \sin\varphi}{r}$$

$$= 2\pi GR^2 \Delta\varrho \, \frac{z}{x^2 + z^2}. \tag{5.10}$$

The profile is symmetrical, and has the property that Δg at $x = z$ is one-half of the maximum value.

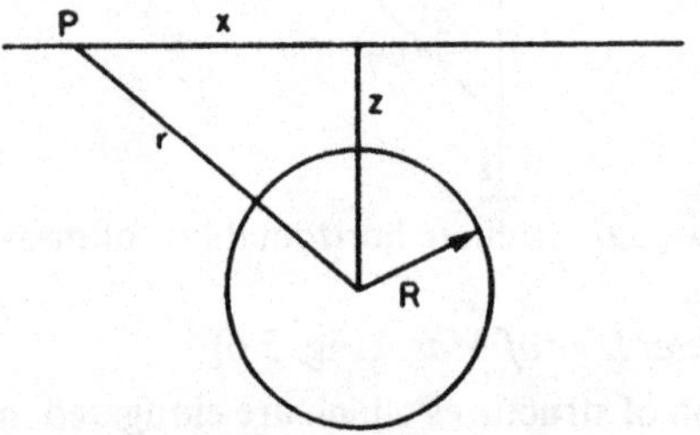

FIG. 5.7.   Horizontal cylinder.

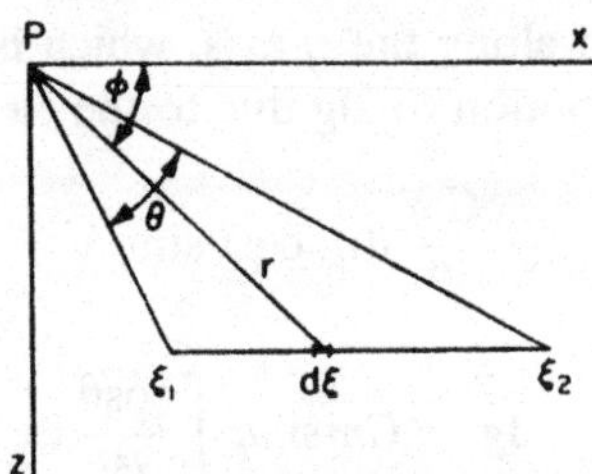

FIG. 5.8.   Two-dimensional horizontal sheet.

Flat Sheet of Infinite Strike Length (Fig. 5.8)

Let the excess mass per unit area be $\Delta\sigma$. The contribution of a ribbon of width $d\xi$ is

$$\frac{2G\Delta\sigma d\xi \sin\varphi}{r}.$$

$$\therefore \; \Delta g = 2G\Delta\sigma \int_{\xi_1}^{\xi_2} \frac{\sin\varphi \, d\xi}{r}$$

or $$= 2G.\Delta\sigma.\theta \qquad (5.11)$$

where θ is the angle subtended by the trace of the sheet in the
x–z plane. We note that this reduces to the Bouguer formula as θ
approaches π.

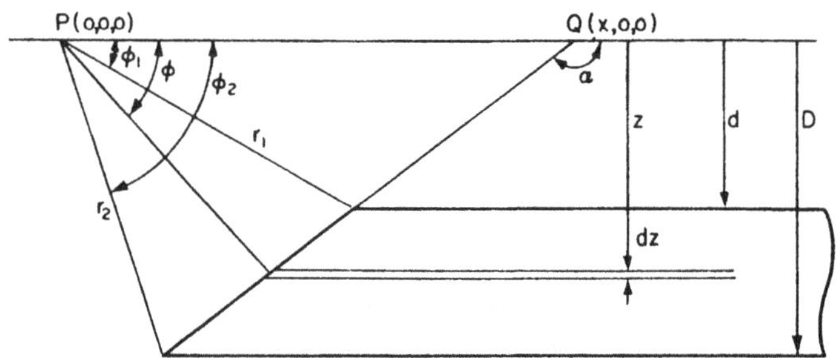

FIG. 5.9.   Sloping edge of a horizontal slab.

Slab with Sloping Edge (Fig. 5.9)

The attraction of a slab, which is infinite in extent in the y- and
also the positive x-direction, can be obtained by superposing the
attraction of sheets of thickness dz. The contribution of the sheet
at depth z is

$$2G\Delta\varrho.\varphi\,dz.$$

$$\therefore \Delta g = 2G\Delta\varrho \int_{d}^{D} \varphi\,dz. \qquad (5.12)$$

After tedious but straightforward integration this leads to

$$\Delta g = 2G\Delta\varrho \left[(D\varphi_2 - d\varphi_1) - \sin a.x\left(\sin a \log_e \frac{r_2}{r_1} \right) + \cos a(\varphi_2 - \varphi_1) \right]. \qquad (5.13)$$

Care is required in the interpretation of the signs of x and
$(\varphi_2 - \varphi_1)$ as P takes up different positions over the slope. If the
slope is very extended compared to its depth and vertical relief,

equation (5.13) reduces to

$$\Delta g = 2G\Delta\varrho\pi(D - x\sin\alpha),\qquad(5.14)$$

which is the Bouguer formula corresponding to the thickness of slab under the point of observation.

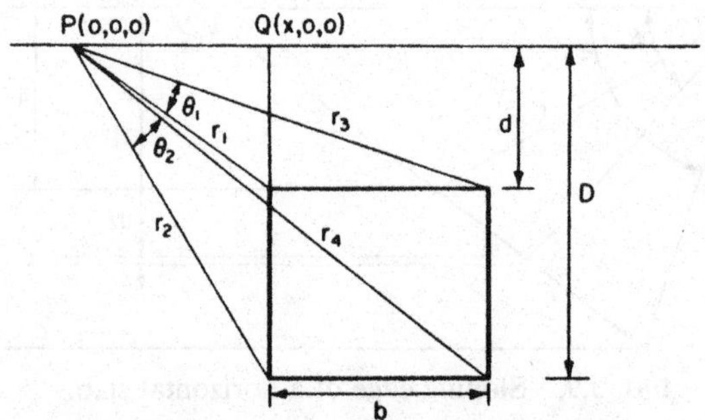

FIG. 5.10. Two-dimensional rectangular prism.

Tabular Bodies or Dikes (Fig. 5.10)

The attraction of these is most conveniently carried out by subtracting the effects of two slopes, displaced horizontally one from the other. For example, the effect of a rectangular prism is obtained by considering two slabs with vertical edges, separated by a horizontal distance b:

$$\Delta g = 2G\Delta\varrho\left[D\theta_2 - d\theta_1 - x\log_e\frac{r_1 r_4}{r_2 r_3} + b\log_e\frac{r_4}{r_3}\right].\qquad(5.15)$$

Two-dimensional Bodies with Irregular Cross-section (Fig. 5.11)

These are best handled by numerical integration through the use of a graticule. Consider first the attraction at P of a segment of a horizontal cylinder:

$$\Delta g = 2G\Delta\varrho\int_{r_1}^{r_2}\int_{\phi_1}^{\phi_2}\frac{\sin\varphi\,r\,d\varphi\,dr}{r}$$

$$= 2G\Delta\varrho(r_2 - r_1)(\cos\varphi_1 - \cos\varphi_2).\qquad(5.16)$$

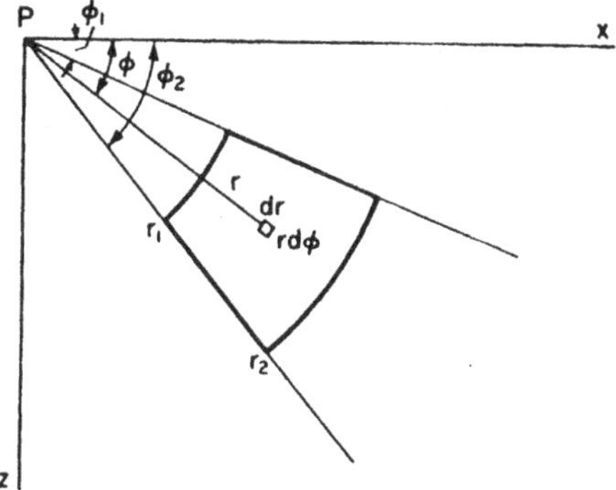

Fig. 5.11. Segment of horizontal cylinder.

If a chart is drawn so that r and $\cos\varphi$ increase in uniform steps through the half-space below P, the contribution at P from all segments will be the same. Convenient steps are 1 cm for r and $0\cdot05$ for $\cos\varphi$. If a scale drawing of the desired cross-section, at a scale of $1:p$, is laid over the graticule, the number of segments contained within the outline can be counted, for any position of the body relative to P. The contribution of each segment to the anomaly, for the segments given, is

$$0\cdot1\ G\varDelta\varrho.p \qquad \text{in gal.}$$

Examples of many other designs for graticules can be found, but the principle of construction and use is the same as that above.

It is convenient, for use in the indirect method of interpretation, to have suites of type profiles computed for various simple shapes. Often, a number of profiles are shown on one diagram, to illustrate the effect of varying one parameter of a body at a time. An example of such a family of curves is shown in Fig. 5.12.

Application of High-speed Computers to Indirect Interpretation

Indirect methods of gravity interpretation have experienced a great increase in usefulness with the availability of high-speed

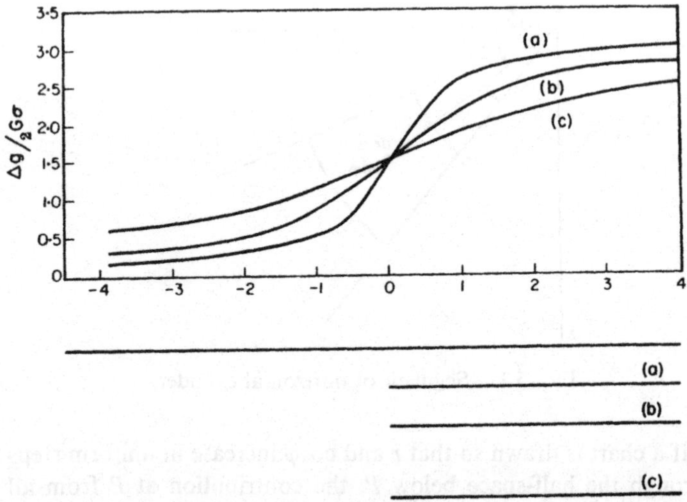

FIG. 5.12. Anomaly profiles across the edge of a mass sheet at different depths ($\frac{1}{2}$, 1 and 2 units of horizontal distance, respectively).

computers. It is now possible to determine rapidly the attraction of irregular shapes, and to observe the changes in the computed anomalies when certain parameters of the assumed bodies are modified.

If computers are to be used in the evaluation of any of the formulae developed above, the expressions must be put into digital form. For example, Talwani and Ewing (1960) developed a digital expression for a solid body, based on a modification of equation (5.7). Any three-dimensional body may be divided by horizontal planes into a series of thin sheets or lamina, the outlines of which may be represented by contours, as in Fig. 5.13.

By means of a theorem of Green, the surface integral of equation (5.7) for any one lamina may be transformed into a line integral around the contour enclosing the lamina, and the contribution to the gravity anomaly may be written

$$\Delta g = G\sigma\left[\int d\psi - \int z/(r^2 + z^2)^{\frac{1}{2}}d\psi\right], \qquad (5.17)$$

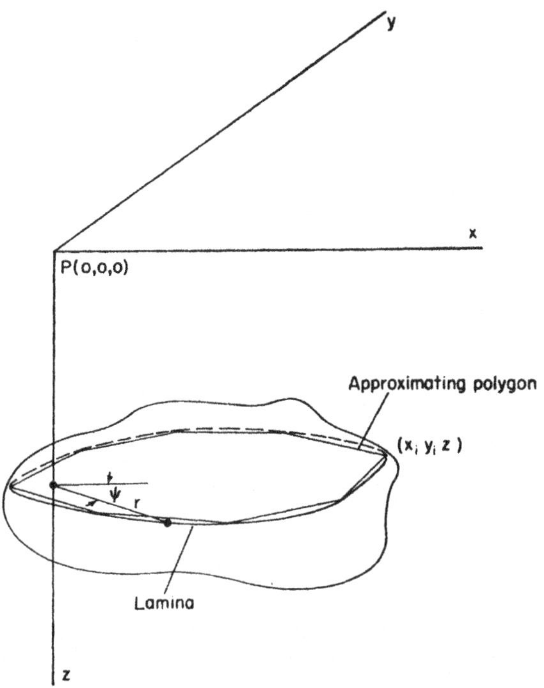

FIG. 5.13. Representation of solid body laminae for high-speed computing of gravity anomaly.

where r, ψ and z are cylindrical coordinates of a point on the contour. For numerical computation, the actual contour may be replaced by a polygon, of which the vertices are (x_i, y_i). Then the expression of equation (5.17) becomes

$$\Delta g = G\sigma \sum_{=1}^{n} \left[W \arccos\left\{ \left(\frac{x_i}{r_i}\right)\left(\frac{x_{i+1}}{r_{i+1}}\right) + \left(\frac{y_i}{r_i}\right)\left(\frac{y_{i+1}}{r_{i+1}}\right) \right\} \right.$$

$$\left. - \arcsin\frac{zq_i S}{(p_i^2+z^2)^{\frac{1}{2}}} + \arcsin\frac{zf_i S}{(p_i^2+z^2)^{\frac{1}{2}}} \right], \qquad (5.18)$$

where $S = +1$ for $p_i > 0$

$\qquad -1$ for $p_i < 0$

$\quad W = +1$ for $m_i > 0$

$\qquad -1$ for $m_i < 0$

$$p_i = \frac{y_i - y_{i+1}}{r_{i.i+1}} \cdot x_i - \frac{x_i - x_{i+1}}{r_{i.i+1}} \cdot y_i$$

$$q_i = \frac{x_i - x_{i+1}}{r_{i.i+1}} \cdot \frac{x_i}{r_i} + \frac{y_i - y_{i+1}}{r_{i.i+1}} \cdot \frac{y_i}{r_i}$$

$$f = \frac{x_i - x_{i+1}}{r_{i.i+1}} \cdot \frac{x_{i+1}}{r_{i+1}} + \frac{y_i - y_{i+1}}{r_{i.i+1}} \cdot \frac{y_{i+1}}{r_{i+1}}$$

$$m_i = \left(\frac{y_i}{r_i}\right)\left(\frac{x_{i+1}}{r_{i+1}}\right) - \left(\frac{y_{i+1}}{r_{i+1}}\right)\left(\frac{x_i}{r_i}\right)$$

$$r_{i,i+1} = [(x_i - x_{i+1})^2 + (y_i - y_{i+1})^2]^{\frac{1}{2}}.$$

The expression for the attraction of the lamina is now completely in digital form, involving only the x- and y-coordinates of the vertices of the approximating polygon. The attraction of the complete three-dimensional body may be obtained by integration over z of the contributions of all laminae, replacing σ by ϱdz, where ϱ as usual is volume density and dz is the interval between contours. In application of the method, the body whose attraction is desired is first represented by horizontal contours, at equal intervals cf z, drawn on squared paper. Sufficient points (x_i, y_i) are then chosen on these contours to represent them to the desired degree of precision, and the coordinates, as read off the diagram, are inserted in the machine programme for the evaluation of equation (5.18). The computer is programmed to determine also the integration over z, so that the complete attraction of the three-dimensional body is obtained. The value of the anomaly at other points is obtained by shifting the origin, adding the desired increments to the values of x_i and y_i.

Direct Methods

Nearly all of the so-called direct methods of interpretation attempt to determine a surface distribution of mass, at a given depth, which is consistent with the observed field. It is shown in Appendix 1 that, for a variable surface density $\sigma(u,v)$ over the u–v plane, the anomaly field very close to this plane is $\Delta g = 2\pi G\sigma(u,v)$. Thus, if the field at depth can be calculated from the field observed at the earth's surface, the mass distribution can be found. Various methods have been suggested for obtaining the field at depth, a procedure known as downward continuation.

Harmonic Analysis

This approach is due to Tsuboi (1938). Consider a plane distribution of mass, at a depth beneath the earth's surface, in which the density σ is a function of x only:

$$\sigma = \sigma_0 \cos px. \tag{5.19}$$

The field on the plane is

$$\Delta g_h = 2\pi G\sigma_0 \cos px \tag{5.20}$$

and it is easy to show that the field at a height h above the plane is

$$\Delta g_0 = 2\pi G\sigma_0 e^{-ph}\cos px = e^{-ph}\,\Delta g_h. \tag{5.21}$$

This suggests that if an observed gravity anomaly, Δg_0, had a purely harmonic form in profile, the field at depth could be obtained through the factor e^{+ph}. However, a profile of any arbitrary shape can be expressed as a sum of Fourier components, each of which can be projected downward by means of the appropriate factor. The field at depth, and therefore the plane distribution of mass, is obtained by the re-combination of the components after projection.

If the anomaly field is a function of two horizontal directions, x and y, it can be expressed as the double Fourier sum

$$\Delta g = \Sigma\Sigma\, c_{pq}\, {\textstyle{\cos \atop \sin}}\, px\, {\textstyle{\cos \atop \sin}}\, qy. \tag{5.22}$$

In this case, a particular component is projected downward

through multiplication by the factor $e^{\sqrt{(p^2+q^2)}h}$, and the interpretation is otherwise the same as for the one-dimensional case. Tsuboi's method was not widely used before high-speed computers were readily available, because of the labour involved in obtaining the Fourier components. However, this analysis presents no difficulty with a computer, and the method is now widely used. Dean (1958) has pointed out that this approach shows clearly the relation of downward projection to a high-pass filtering operation. Short wavelength, high "frequency" anomalies, are represented by large values of p, or p and q. For projection to a given depth h, these features are accentuated by the factor e^{+ph} or $e^{+\sqrt{(p^2+q^2)}h}$. In fact, these terms may lead to components of σ which are so large in amplitude as to be completely unrealistic. This may mean that the solution is sought at too great a depth; that is, that too large a value of h has been chosen, and the true source of the anomaly is shallower. A limit may therefore be placed on the maximum acceptable depth for the mass distribution. On the other hand, the oscillation of σ may be the result of lack of smoothness in the observed profile. Random errors in the observation or reduction of g lead to Fourier components whose half-wavelength is two units of station spacing, which is normally short compared to the entire profile. It is desirable that any anomaly profile or map be smoothed before downward projection is undertaken. Dean (1958) describes methods of smoothing, and also the construction of computation programmes, for the projection.

Taylor Expansion

The anomaly field at depth may also be determined from the field and its vertical derivatives on the surface (Evjen, 1936). We take the z-axis vertically downward, with the earth's surface as the x–y plane. Then the field at depth h is given by

$$\Delta g(x,y,h) = \Delta g(x,y,0) + \left[\frac{\partial(\Delta g)}{\partial z}\right]_{x,y,0} h + \left[\frac{\partial^2(\Delta g)}{\partial z^2}\right]_{x,y,0} \frac{h^2}{2!} + \cdots$$

$$(5.23)$$

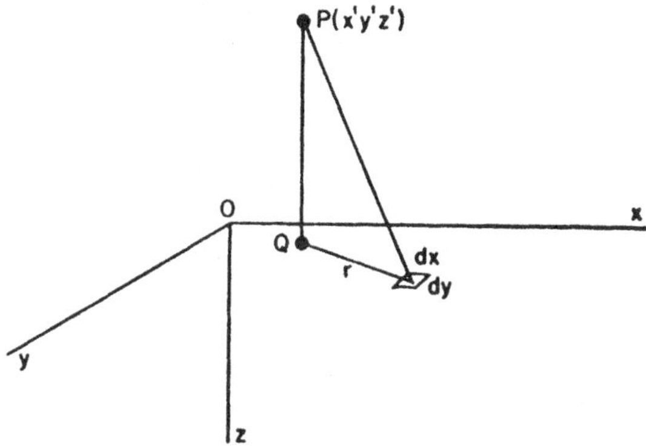

FIG. 5.14. Quantities involved in surface integration of the gravity field.

Downward projection therefore depends on the calculations of the derivatives, and this is possible if the field is given in sufficient detail on the earth's surface (Peters, 1949). The field at all points on and above this surface will be unchanged if the actual cause of the anomaly is removed, and a surface distribution $\Delta g(x,y,0)/2\pi G$ is spread over the x–y plane (Appendix 1). Then, at a point P above the plane (Fig. 5.14), the contribution to the gravity anomaly of the surface element $\mathrm{d}x\mathrm{d}y$ is

$$\frac{-z\Delta g(x,y,0)\ \mathrm{d}x\mathrm{d}y}{2\pi\,[(x'-x)^2+(y'-y)+z'^2]^{3/2}}$$

and the complete anomaly at P is

$$\Delta g(x'y'z') = \frac{-z'}{2\pi}\int\limits_{-\infty}^{\infty}\int\limits_{-\infty}^{\infty}\frac{\Delta g(x,y,0)\mathrm{d}x\mathrm{d}y}{[(x'-x)^2+(y'-y)^2+(z')^2]^{3/2}}. \tag{5.24}$$

Equation (5.24) gives the anomaly field at P in terms of a surface integral over the x–y plane. Derivatives of the field can now be obtained by differentiation with respect to z through the integral. For example,

$$\left[\frac{\partial \Delta g}{\partial z}\right]_{x'y'z'} = -\frac{1}{2\pi} \int_{-\infty}^{\infty} \int_{-\infty}^{\infty} \Delta g(x,y,0) \times$$

$$\times \left[\frac{(x'-x)^2+(y'-y)^2-2z'^2}{[(x'-x)^2+(y'-y)^2+2z'^2]^5}\right] dxdy. \quad (5.25)$$

Further differentiation changes only the weighting function contained under the integration sign. The evaluation of integrals of the type contained in equation (5.25) is most conveniently carried out by changing to polar coordinates (r,θ) about the point Q (Fig. 5.14). For example,

$$\left[\frac{\partial \Delta g}{\partial z}\right]_{0,0,z} = -\frac{1}{2\pi} \int_{r=0}^{\infty} \int_{\theta=0}^{2\pi} \Delta g(r,\theta,0)\left[\frac{r^2-2z'^2}{(r^2+z^2)^{5/2}}\right] rdrd\theta. \quad (5.26)$$

If we let the average value of Δg around a circle of radius r be $\overline{\Delta g(r)}$, we have

$$\overline{\Delta g(r)} = \frac{1}{2\pi} \int_{0}^{2\pi} \Delta g(r,\theta,0)d\theta \quad (5.27)$$

and

$$\left[\frac{\partial \Delta g}{\partial z}\right]_{0,0,z'} = \int_{0}^{\infty} \overline{\Delta g(r)} \left[\frac{r^2-2z'^2}{(r^2+z'^2)^{5/2}}\right] rdr. \quad (5.28)$$

In practice, the mean value of the anomaly will be computed for circles of different radii, 0, r_1, r_2, etc., about Q, and the desired quantity is obtained from the sum

$$\left[\frac{\partial \Delta g}{\partial z}\right]_{0,0,z'} = -\left[\frac{\overline{\Delta g_0}+\overline{\Delta g_{r_1}}}{2} \int_{0}^{r_1} \left(\frac{r^2-2z'^2}{(r^2+z'^2)^{5/2}}\right) rdr \right.$$

$$\left. + \frac{\overline{\Delta g_{r_1}}+\overline{\Delta g_{r_2}}}{2} \int_{r_1}^{r_2} \left(\frac{r^2-2z'^2}{(r^2+z'^2)^{5/2}}\right) rdr +\ldots \right]. \quad (5.29)$$

This leads to

$$\left[\frac{\partial \Delta g}{\partial z}\right]_{0,0,z'} = \frac{\overline{\Delta g_0}}{2}\left[\frac{1}{(r_1^2 + z'^2)^{1/2}} - \frac{z'^2}{(r_1^2 + z'^2)^{3/2}}\right]$$

$$+ \frac{\overline{\Delta g_1}}{2}\left[\frac{1}{(r_2^2 + z'^2)^{1/2}} - \frac{z'^2}{(r_2^2 + z'^2)^{3/2}}\right] + \cdots \quad (5.30)$$

As the value of the vertical derivatives of gravity is required on the earth's surface, z' is now allowed to approach zero. Then

$$\left[\frac{\partial \Delta g}{\partial z}\right]_{0,0,0} = \frac{\overline{\Delta g_0}}{2}\left[\frac{1}{r_1}\right] + \frac{\overline{\Delta g_1}}{2}\left[\frac{1}{r_2}\right] + \cdots \quad (5.31)$$

The coefficients in the series for $\partial \Delta g/\partial z$, as well as those in the corresponding series for all higher derivatives, reduce to numerical constants when the radii of the successive circles have been chosen. The procedure for computation is straightforward, provided the number and distribution of gravity stations are such that the average values of anomaly, $\overline{\Delta g}$, can be obtained. If the stations are located on a uniform grid, the values of r can usually be chosen to make the circles pass through several points of observation. Otherwise, the anomalies will have to be contoured first, to permit values on the circles to be estimated. Errors in the anomalies, or in the estimating of circle averages, will be accentuated, particularly in the higher derivatives. The number of terms which must be evaluated in the expansions for the derivatives, and the number of derivatives required in the expression for gravity at depth, will depend on the complexity of the field in question and on the precision which is required. Further details on the computation are given by Evjen (1936) and Peters (1949).

Direct methods of interpretation are most useful in those cases in which a good distribution of stations, of high accuracy, is available. In most methods, such as those outlined above, the interpretation consists of the determination of a mass distribution over a horizontal plane, although Grant (1952) has described a method which is not restricted to sheet-like structures.

CHAPTER 6

Gravity Anomalies
and the Interior of the Earth

The Nature of the Earth's Interior

We wish to discuss first the broadest aspects of gravity anomalies, with special reference to the degree to which isostatic compensation is achieved. It is necessary to have, as a background for this, a brief look at our knowledge of the earth's interior derived from sources other than gravity.

The most detailed knowledge of the interior is obtained from seismological observations: both of the distribution and nature of earthquakes, and of the velocities of elastic waves which travel through the earth. It is known that at a depth of a few tens of kilometres below sea level there is an abrupt increase in the velocities of both longitudinal and transverse waves. The surface at which this change takes place is called the Mohorovicic discontinuity, and is taken to mark the lower boundary of the crust. The elastic properties of the sub-crust or mantle suggest that it is a material containing more iron and magnesium, and less calcium, aluminium and free silica, than the crustal rocks. It may resemble the mineral olivine, which is an iron-magnesium silicate occurring at the earth's surface, in rather limited areas of very basic rock. The density immediately beneath the Mohorovicic discontinuity is probably of the order of $3 \cdot 2$ g/cm^3, so that there must be a considerable discontinuity in density, as well as elastic wave velocity, at that surface.

Studies of seismic waves from explosions and near-earthquakes have shown that the velocity usually increases with depth in the crust. In many areas, the crust appears to be layered, with each layer having characteristic properties. An investigation of the longitudinal and transverse velocities for two layers indicated in the crust of Europe led Jeffreys (1952) to suggest that the materials in the upper and lower layer were respectively acidic and intermediate in properties. The terms "granitic layer" and "basaltic layer" have become widely used, but more recent work has shown that the velocity conditions in the crust can vary greatly within remarkably small horizontal distances. In particular, the occurrence of granite in the upper part of the crust will be discussed in the following chapter.

The mantle extends in depth to 2900 km, which is the location of the boundary of the central core. Seismological investigations show that the core is liquid, and that there is a further discontinuity in density at its boundary. Most explanations of the origin of the earth's magnetic field require that the core material be a relatively good electrical conductor; it may be molten iron, or a high-pressure, metallic phase of silicates.

Earthquakes, which represent the sudden failure of material under stress, occur at depths as great as 750 km, although the frequency decreases with depth. There is thus a suggestion that the response of the mantle to stress below this depth is different from that above. It is known also that the elastic wave velocities and other properties change abruptly, possibly discontinuously, at about the same depth, and a phase change, under pressure, of the mantle silicates has been suggested as a cause.

A problem of prime importance in both geophysics and geology is the determination of the mechanism responsible for the earth's surface features. The rocks exposed in mountain ranges and ancient shields show evidence of intense deformation. Some system of horizontal forces is required to explain this deformation, but there are many theories as to the origin of the forces. Nearly all of these, however, depend on some assumptions regarding the thermal state of the earth's interior. For example, the deformation of the crust has been attributed in some theories to contraction of the earth during cooling, and in others to the influence of convection currents in the mantle. The recently extended evidence for continental drift, on the basis of palaeomagnetic observations, has suggested to many geophysicists that convection currents in the mantle may well play an important role in producing major dislocations at the surface.

The study of the major features of the earth's gravitational field can throw light on these theories by indicating the nature of the mantle material, and the form of crustal deformations. Evidence on the rheological nature of the mantle is available from sources other than the static field; both the tidal variations of gravity (Chapter 9) and certain characteristics of the earth's rotation (Jeffreys, 1952, chapter 7; Munk and MacDonald, 1960) also provide information. But gravity anomalies, particularly those which indicate the presence of uncompensated masses on the earth, may show the behaviour of the mantle under stresses of long duration. Alternatively, they may indicate the presence of convection currents in the mantle.

Low Order Harmonics in the Gravitational Field

The low order harmonics, which produce the undulations in the geoid shown in Fig. 4.7, represent the gravity anomalies of broadest extent. However, the presence of free-air gravity anomalies is not in itself proof of uncompensated mass. A topographic feature which is completely compensated at some depth will still produce a free-air anomaly. It is the relation between the spherical harmonics in the expansion for free-air gravity and those in the expansion for the topography which indicate the degree of compensation.

Consider a spherical harmonic term of order n in the topography. If the coefficient of this term is E_n, and the density of the crust is ϱ, there is an equivalent mass per unit area $\sigma_n = \varrho E_n S_n$ spread over the spheroid, where S_n is a spherical harmonic of order n. For perfect compensation by the Airy mechanism at depth H, there will be an equal mass spread over the sphere of radius $(a - H)$, the amplitude of the compensating surface density σ_c being

$$\sigma_c = -\sigma_n \frac{a^2}{(a-H)^2}. \tag{6.1}$$

However, this distribution of mass does not make the surface of compensation an equipotential of the field. Since this surface should represent the outermost layer in which hydrostatic conditions exist, it is reasonable to assume that it should be a surface of constant potential. Following Jung (1952), we shall vary the compensating masses somewhat from the strict Airy conditions to provide this equality of potential over the sphere of radius $(a - H)$. The internal potential of the topography on a sphere of radius $(a - H)$ is (Appendix 1)

$$\frac{4\pi G}{2n+1}\,\sigma_n\,S_n\,\frac{(a-H)^n}{a^{n-1}}$$

while that of the compensating masses is

$$\frac{4\pi G}{2n+1}\,\sigma_c\,S_n\,(a-H).$$

Therefore, if the sphere of radius $(a - H)$ is to be an equipotential

$$\frac{\sigma_c}{\sigma_n} = -\left(1-\frac{H}{a}\right)^{n-1} \tag{6.2}$$

The net external potential of the topography and the compensation is then

$$V_n = \frac{4\pi G S_n}{2n+1}\left[\sigma_n\frac{a^{n+2}}{r^{n+1}} - \left(1-\frac{H}{a}\right)^{n-1}\sigma_n\frac{(a-H)^{n+2}}{r^{n+1}}\right] \tag{6.3}$$

The free-air gravity anomaly is obtained by differentiation of (6.3), with the indirect effect due to the undulation of the geoid, $2V/a$, subtracted. We neglect a small effect due to attraction of material between the spheroid and geoid. Then,

$$\Delta g_n = 4\pi G\frac{n-1}{2n+1}\left[1-\left(1-\frac{H}{a}\right)^{2n+1}\right]\sigma_n S_n. \tag{6.4}$$

The ratio of corresponding terms in the expansion for topography and free-air gravity is then

$$\frac{\Delta g_n}{E_n} = 4\pi G \frac{n-1}{2n+1}\left[1-\left(1-\frac{H}{a}\right)^{2n+1}\right]\varrho. \qquad (6.5)$$

As an example, for $n = 2$, the amplitude of free-air gravity anomaly for 1 km of topography of normal density compensated at a depth of 50 km is $1\cdot4$ mgal; for $n = 10$, it is $11\cdot8$ mgal. If the topography were uncompensated the anomaly would be 112 mgal. The reduction of the free-air anomaly, in the compensated case, with increasing wavelength is to be expected, as the opposite effects of direct attraction and compensation more nearly annul each other for the low harmonics. If there are harmonics in the expansion of free-air gravity which are unrelated to corresponding harmonics in the topography, they must be caused by uncompensated masses.

Kaula (1959), having obtained the spherical harmonics for both free-air gravity and topography up to order 8, used an equation similar to (6.5) to determine H, the indicated depth of compensation. Excluding second order spherical harmonics (which are sensitive to the assumed ellipticity of the spheroid), he obtained an average depth of 37 km, which agrees well with investigations of Airy isostasy in specific areas by other approaches. However, many individual harmonic terms are not well satisfied by this depth. In some cases, the terms in the topography and gravity are of opposite sign, which can only indicate uncompensated masses within the earth, unrelated to topography. The recent analyses of the free-air field, such as that of Kaula, have shown a smaller longitudinal variation, especially around the equator, than the analysis of Jeffreys (1943). Nevertheless, it still appears that there are anomalies, indicative of uncompensated masses, of amplitude of the order of 10 mgal, with wavelengths corresponding to the first eight harmonics. The low-harmonic anomalies are of particular interest because they may be related to conditions at relatively great depth within the earth.

Relation between Uncompensated Masses and Stress

It is important to realize that perfect isostatic compensation, if it existed, would provide a clue as to the nature of the mantle, rather than the crust. Perfect isostasy implies a state of hydrostatic stress below the depth of compensation; conversely, departures from this state imply a strength in shear on the part of the mantle material. The maximum stress-difference which the rocks of the crust can support is known to be of the order of 1×10^9 dynes/cm^2, and stress-differences of this magnitude occur in the crust, even beneath compensated topographic features. If crustal strength were absent, the features themselves would disappear.

Jeffreys (1952) has investigated in detail the stress-difference (i.e. difference between maximum and minimum principal stress) produced in an elastic earth by different types of surface loads. The case of the load which varies harmonically with distance in one dimension on a flat earth can be treated without difficulty by the usual equations of elasticity, and serves to indicate the general result. The stress-difference reaches a maximum value, equal to two-thirds of the load, at a depth of $1/2\pi$ times the wavelength of the loading. For example, consider an uncompensated mass, represented by a wavelength of 12,000 km (a spherical harmonic of order 3) which produces a free-air anomaly of 10 mgal. The equivalent amplitude of surface loading is $2 \cdot 3 \times 10^7$ dynes/cm^2, and the simple solution above suggests that stress-differences of $1 \cdot 7 \times 10^7$ dynes/cm^2 would be produced, at a depth, in a uniform earth, of approximately 2000 km. If the condition is made that the mantle material beneath a certain depth, say 600 km, has no strength, the stress-difference in the upper mantle becomes considerably greater. It is by arguments similar to these that Jeffreys concluded that the upper 600 km of the mantle must be able to withstand long-term stress-differences of $3 \cdot 3 \times 10^8$ dynes/cm^2. In other words, the material would require a strength of about one-third that of rocks in the crust.

Gravity Anomalies and Convection Currents

The above considerations assume that the mantle may be treated as an elastic body. If it is assumed that it is a viscous fluid capable of convection, some other explanation must be found for the long wavelength gravity anomalies. A natural question that may be asked is what effect any assumed pattern of convection currents would have on gravity at the surface.

Convection (Rayleigh, 1916; Pekeris, 1935; Hales, 1935) consists of transfer of heat by motion of material, hot material rising along certain paths, and cooler material sinking along others. Most of the theories of convection proposed for the earth's mantle have suggested the presence of relatively few, large-dimension convection cells. The motion of mantle material is assumed to produce up- and down-warps, and horizontal dislocations, of the crust. Up-warp and tearing apart would be associated with areas over rising currents, while continental blocks would tend to move toward sinking currents.

The calculation of the surface anomaly field to be expected from even a simple pattern of convection currents is complicated by the combination of density variations with temperature, and the distortion of the surface. Suppose that a zonal perturbation of temperature is assumed to exist within the mantle, of the form $f(r)P_n(\cos\theta)$, where r is radial distance, θ the latitude, and P_n a spherical harmonic of order n. Then the perturbation of density at any point will be $-a\varrho_0 f(r)P_n(\cos\theta)$, where a is the volume coefficient of thermal expansion, and ϱ_0 is the undisturbed density. The temperature disturbance creates a potential U_1, which satisfies Poisson's equation

$$\nabla^2 U_1 = 4\pi G a \varrho_0 f(r)P_n(\cos\theta) \qquad (6.6)$$

inside the earth, and

$$\nabla^2 U_1 = 0 \qquad (6.7)$$

at external points.

Solutions to equations (6.6) and (6.7) can be found, if the function $f(r)$ can be chosen in simple form. The free-air anomaly in gravity, with allowance for the indirect effect, is then

$$\Delta g_1 = -\left(\frac{\partial U_1}{\partial r}\right)_{r=a} - \frac{2U_1}{a}, \qquad (6.8)$$

where a is the earth's radius. The gravity anomalies due to this cause will be positive over sinking currents, where the temperature is decreased by the perturbation, and negative over rising currents. However, if there is a warping of any surface of discontinuity in density within the earth, an additional contribution to the gravity anomaly, of opposite sign, is produced. For example, suppose that the base of the crust is warped by an amount $uP_n(\cos\theta)$, and that at this level the discontinuity in density is $\Delta\varrho$. Then there is a distribution of anomalous mass, of surface density $\Delta\varrho uP_n(\cos\theta)$, spread over the sphere of radius $a - H$, where H is the thickness of the crust. The potential of this distribution is then (Appendix 1)

$$U_2 = \frac{4\pi G}{2n+1}\left(\frac{u.\Delta\varrho(a-H)^{n+2}}{r^{n+1}}\right)P_n(\cos\theta). \qquad (6.9)$$

The free-air gravity anomaly will be given by an equation analogous to equation (6.8). The difficulty in evaluating the net gravity anomaly comes in estimating the relative contribution of U_1 and U_2. Pekeris (1935) investigated zonal convection patterns in the mantle beneath a thin crust, and assumed that at each level the uplift was such that the weight of uplifted material equalled the normal stress produced by convection. In this case, the gravity anomalies for convection cells of order 2 were found to be small (less than 20 mgal), because of the near-cancellation of the two effects, but they were positive over areas of uplift. If convection were assumed to exist in the mantle only below the maximum depth of earthquakes, say 800 km, the situation could be very different. Within such a thick shell the uplift might be relatively much less, even though the tangential stresses set up by the currents were still effective in tectonic activity at the surface. If this were the case, the effect of the temperature distribution would predominate, and negative anomalies would be expected over areas of rising currents. In terms of warping of the geoid (Fig. 4.7) outward warps would then correspond to sinking currents. It is not immediately obvious that the actual geoidal warpings correspond to any simple low harmonic. However, it is possible to

D

estimate very roughly the temperature variations in the mantle which would be required to produce a gravity anomaly of given harmonic order. If a constant temperature difference T is assumed to exist between the rising and falling currents, over the interval from the core boundary to a depth of 800 km, equation (6.6) leads to a free-air gravity anomaly of $2\,T$ mgal, for $n = 2$ and a value of α of $60 \times 10^{-6}\,°C^{-1}$. Since the low-order free-air gravity anomalies are 20 mgal or less, temperature differences of only a very few °C would be required to explain them, on this model.

It is rather ironic that the low-order gravity anomalies could be indicative either of long-term strength in the earth's mantle or of convection currents, as these two conditions are rheologically incompatible, and the geophysicist must choose between them.

Regional Studies of Isostasy

The very broad, if low amplitude, departures from compensation discussed in the preceding sections are currently of great interest, but they should not obscure the fact that over structures of continental size isostasy is at least a very good first approximation. This is indicated by numerous studies (Heiskanen and Vening Meinesz, 1958, Chapter 7) that are available for specific regions. Comparisons between Bouguer, free-air and isostatic anomalies almost invariably show that the latter are closest to zero. On the other hand, in every region there is a scatter in the isostatic anomalies which is not reduced by any variation in parameters. Table 6.1 shows the comparison for mountain stations in the United States (chiefly in the Cordillera), and the Alps, two young mountain ranges, and for stations at sea.

The large excursions of the Bouguer anomalies, from positive values over the oceans to negative values over the continents, are well illustrated. The Airy–Heiskanen isostatic anomalies are computed for different assumed values of crustal thickness at sea level. The effect of making the reduction with different thicknesses is considerable, as shown by the alpine stations.

However, because of the ambiguity of gravity interpretation, comparisons of this type do not prove that a particular mechanism

TABLE 6.1

	Mean anomalies (mgal)				
	Free-air	Bouguer	Hayford 113·7 (km)	Airy–Heiskanen 20 (km)	40 (km)
United States (11 mountain stations)	+75	−100	+15		+13
Alps (23 stations)	−14	−115		+2	−13
23 sea stations	− 6	+269		−1	

is correct, nor do they provide the material in a form convenient for investigating in detail the most probable distribution of the compensating masses. As seismological measurements give a detailed picture of the outer layers of the earth, isostatic investigations should be correlated with these wherever possible.

The Airy Crust and the Seismological Crust

The Mohorovicic discontinuity, which defines the base of the seismological crust, represents the chief discontinuity in elastic wave velocity in the outer part of the earth. Measurements of the velocity of longitudinal and transverse elastic waves on rock samples show that density and velocity tend to vary together, and it is reasonable, therefore, to expect an abrupt change in density at Mohorovicic discontinuity.

The expressions for the elastic wave velocities can be written

$$V_p = \sqrt{\frac{K + \frac{4}{3}\mu}{\varrho}}$$

$$v_s = \sqrt{\frac{\mu}{\varrho}}, \tag{6.10}$$

where K is the bulk modulus, μ the shear modulus, and ϱ the density. If the elastic moduli remained constant, both velocities would vary inversely as the square root of density. In fact, however, the

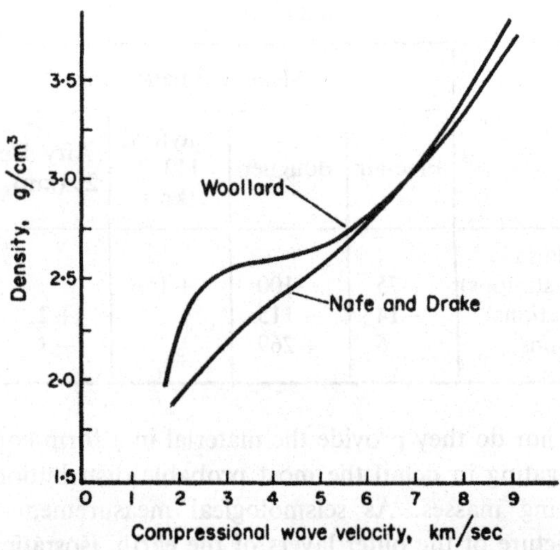

FIG. 6.1. Variation of density with compressional wave velocity.

moduli of rocks increase with decreasing free silica and increasing ferromagnesian mineral content. Measurements by Nafe and Drake (1957) show that velocity and density increase together, as indicated in Fig. 6.1, and the Mohorovicic discontinuity must be a surface at which density increases by perhaps $0\cdot4$ g/cm³. If Airy isostasy is effective, we should therefore expect a major portion of the compensation to be accomplished at that interface. That this is so in a general way is indicated by Fig. 6.2, in which a number of seismological determinations of crustal thickness are shown. The diagram indicates the topographic height or ocean depth for the area of the determination, and, where available, the Bouguer and isostatic gravity anomalies. There is no question that the Mohorovicic discontinuity mirrors the topography, as regards the differences between oceans and continents. Within the continents, the greatest crustal thickness is observed beneath the Andes mountains of South America, and the Academy of Science mountains of the U.S.S.R. The crust is thinnest beneath the deepest oceans, and tends to thicken beneath mid-ocean ridges and

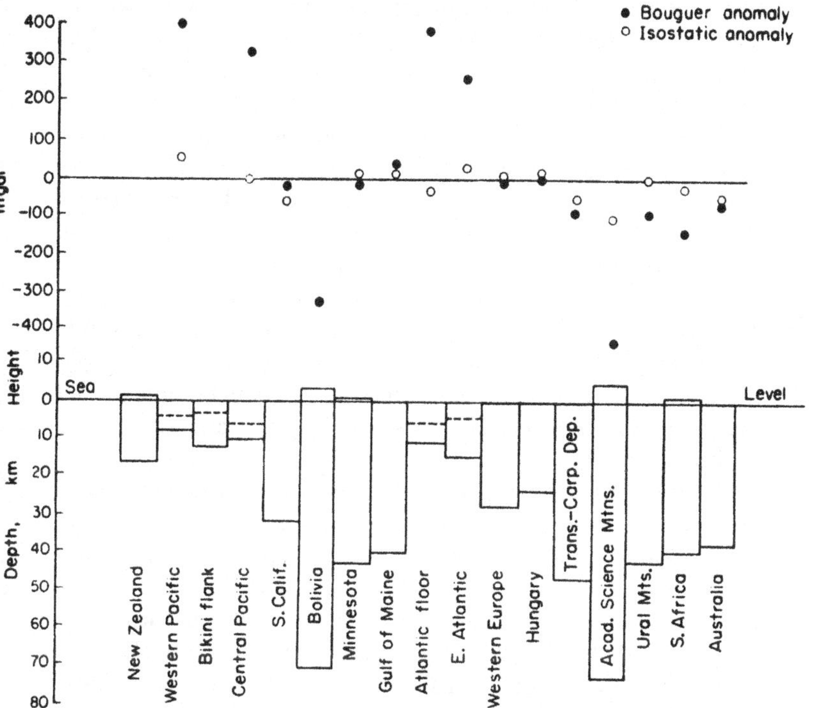

Fig. 6.2. Crustal sections determined by refraction seismology over oceans and continents, showing height (or depth) of land surface, depth to Mohorovicic discontinuity, and gravity anomalies.

oceanic islands. We conclude, therefore, that on a world-wide scale the seismological crust does indeed play the role of the crust visualized by Airy.

The gravity anomalies plotted on Fig. 6.2 show the striking excursions of the Bouguer anomalies to large positive values over the oceans, and to large negative values in mountainous areas of the continents. Because the attraction of the topography, or the deficiency of attraction of the ocean, is corrected for in computing these anomalies, their values reflect the compensation, in particular, the variations in thickness of the lower-density crust. The isostatic anomalies, on the other hand, remain much closer to zero over both oceans and continents, indicating that the crustal thickness variations are close to those required for compensation. However, there are some regions below which the crustal thickness is not appropriate to the topographic mass, and there is either over- or under-compensation. Referring again to Fig. 6.2, the Trans-Carpathian depression is a region of low relief, but the crust is thicker than normal for a continental area. The Bouguer anomaly is more negative than for other regions of comparable height, and the isostatic anomaly is also negative. Conversely, the crust beneath the western Pacific Ocean is thinner than normal for the particular depth of water, and the isostatic anomaly is positive.

Because Airy isostatic anomalies were not available for all areas shown in Fig. 6.2, some of the values plotted are Pratt–Hayford anomalies. Also, in some cases these represent average values over a 1° "square", rather than determinations at the exact site of the seismological measurements. However, the general conclusion that the isostatic anomalies remain much closer to zero than the Bouguer anomalies is still valid. Generally, Pratt–Hayford anomalies are found to be least when the depth of compensation for this case is taken to be slightly over 100 km. There is no evidence for a discontinuity in density at this depth. Uniform compensation distributed in depth, as in the Pratt model, would produce an effect on gravity very nearly equivalent to that of Airy-type compensation at half the depth. Hence, minimum isostatic anomalies found by the two methods are comparable in magnitude.

Elevation and Crustal Thickness

The general world-wide relationship, found in the previous section, between topographic height and crustal thickness is not confirmed in detail when a number of crustal sections from one continent are studied. Figure 6.3 illustrates the results of seis-

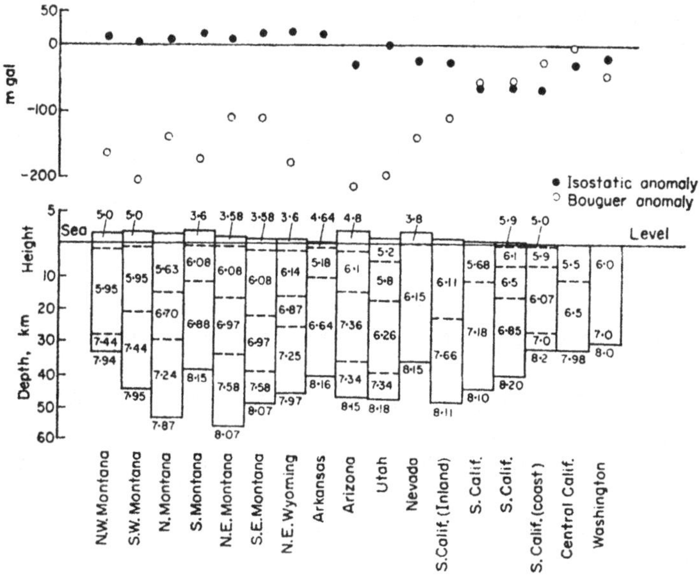

FIG. 6.3.   Crustal sections determined by refraction seismology in the western United States. Numbers indicate compressional wave velocity in km/sec.

mological measurements of crustal thickness for the western United States (Steinhart and Meyer, 1961). Certain areas of high relief, such as western Montana and Nevada, do not have as thick a crust as areas of lesser elevation. It should be noted that not all of the sections shown in Fig. 6.3 represent strictly independent seismological measurements, as a section is shown for each end of each available reversed seismic profile. However, there is good evidence that the topographic height cannot be used by itself to predict crustal thickness. Figure 6.3 also shows Bouguer and isostatic (Pratt–Hayford) anomalies for each section. In some cases, the Bouguer anomaly mirrors the topography even where the base of the crust does not, for example, in the case of the Montana sections. For these, the isostatic anomaly is uniformly small, indicating that compensation is achieved, although apparently not by the Airy mechanism alone. In the case of the Nevada section, the crust is much thinner than would be predicted from the height, but the isostatic anomaly is actually negative. Some of the California sections indicate considerable over-compensation, on the basis of the negative isostatic anomalies, although the crust is not abnormally thick there. Obviously, there is an influence present not directly related to crustal thickness. The velocities for the crustal layers and the sub-crust, shown on Fig. 6.3, suggest what this may be. The properties of the crust are obviously variable from place to place. In some cases, such as Nevada, it appears to consist of a single layer, below the near-surface material, of velocity 6·15 km/sec. In other sections, layers with velocity as high as 7·58 km/sec are indicated. On the basis of the density–velocity curve, Fig. 6.1, the mean density of the crust must vary considerably between the sections shown. The material of the upper mantle, below the Mohorovicic discontinuity, must also have properties which vary with location, as its velocity is shown as ranging from 7·87 to 8·20 km/sec. In those areas in which compensation is maintained even where the base of the crust does not mirror the topography, these property variations must play a role. There is thus a suggestion of the mechanism of Pratt.

It is instructive to compare the total mass per unit area in the different sections with that of some "standard" continental section. Worzel and Shurbet (1955), by comparing gravity measurements and seismic refraction measurements made along the east coast of North America, suggested standard sections for continental areas at sea level and for oceanic areas. The densities and thicknesses for these sections are shown in Fig. 6.4. Woollard (1959) suggested that the mean density of the continental crust might be closer to 2·87 than 2·84 g/cm³. Taking a standard crust as one of density 2·87 g/cm³ and thickness 33 km, it is possible to compute the mass excess or deficiency below sea level from the seismic section and the density–velocity curve, Fig. 6.1. If an elevated region is compensated, the mass deficiency below sea level must equal the mass of the topography above sea

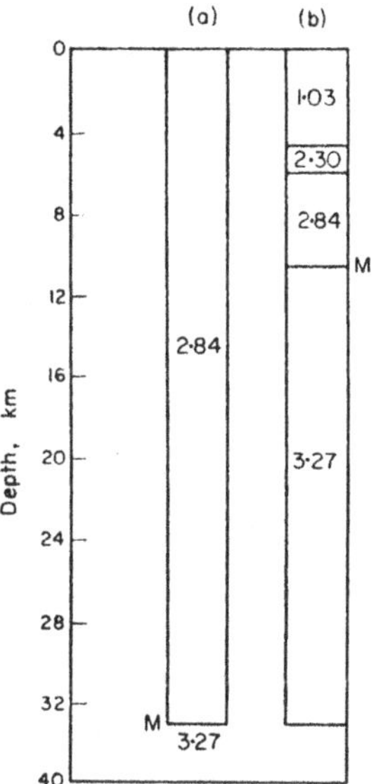

FIG. 6.4.   Standard continental (a) and oceanic sections as proposed
by Worzel and Shurbet. Numbers are densities in g/cm³.

level. A comparison for four of the United States sections is given
in Table 6.2.

The table shows also the mass imbalance which, in the form of an
infinite sheet, would produce the observed isostatic anomalies. It
is obvious that the sections as determined by seismic refraction
observations do not provide the required compensation. For example, in the case of the crust beneath north-eastern Montana, the
effect of crustal thickening is almost nullified by the large proportion of high velocity, high density crustal rock. There are, of
course, a number of difficulties in this type of comparison, including the uncertainty of seismic interpretation for intermediate
layers, and the probable error of the curve giving the relation
between density and velocity. However, the crustal imbalances
shown in Table 6.2 are so much greater than the observed depar-

TABLE 6.2

	Southern California	Nevada	North-eastern Montana	North-western Montana
Height (m)	100	1830	900	1500
Density above sea level (g/cm³)	2·66	2·61	2·60	2·66
Mass excess (10^5 g/cm²)	0·266	4·80	2·34	4·00
Total crustal thickness (km)	32	36	57	35
Mass deficiency in crust, relative to standard section (10^5 g/cm²)	1·20	1·65	0·40	0·88
Apparent mass excess in crust (10^5 g/cm²)	−0·93	3·15	1·94	3·12
Regional isostatic anomaly (mgal)	−20	−10	+20	+10
Mass excess indicated by anomaly (10^5 g/cm²)	−0·48	−0·24	+0·48	+0·24

tures from compensation that these uncertainties can hardly be responsible. Almost certainly some compensation must be provided by density variations below the Mohorovicic discontinuity. This is suggested by the trend in mantle velocity to lower values below the mountainous areas. A difference in density of 0·2 g/cm³, existing to 10 or 15 km below the base of the crust, would provide the additional compensation required.

Variations in density in the upper mantle, particularly beneath mountains, are to be expected if crust and mantle became mixed during mountain building (Cook, 1962). In other words, the boundary now mapped as the Mohorovicic discontinuity, and shown in Fig. 6.3, may not represent the original base of the crust. Under the action of horizontal forces, a section of the crust may not only shorten and thicken, but may to a considerable extent become mixed with the underlying mantle. Differentiation, and the addition of mineralizing solutions, could be expected to produce an upper crust of typically granitic properties, with relatively low density and velocity. The lower part of the new mountain root would consist of material ranging from basalt to ultrabasic rock, with density and velocity somewhat lower than those in the undisturbed mantle. This material may well grade into the underlying mantle in such a way that seismological observations cannot determine the lower boundary of the mixed crust. If this is the case, the designation of a particular discontinuity below mountainous areas as the Mohorovicic discontinuity is rather arbitrary, particularly if no velocity as high as 8·0 km/sec is observed.

A more complete explanation of the mechanism of compensation must await further investigations of the crust, both seismological and gravimetric. In particular, the relation between density and velocity requires confirmation, and the regional variations in upper mantle properties must be established. With the present information it can be said that density variations both above and below the seismologically determined base of the crust appear to be required to supply the compensation in those areas where the Airy mechanism alone is not sufficient.

It should be pointed out that other mechanisms of isostatic compensation have been suggested by various workers, and that these involve modifications of the usual Airy and Pratt conditions. Vening Meinesz (1931, 1941) assumed that a load placed on a floating crust would cause it to bend as an elastic plate, and that the compensation of the load would not be concentrated beneath it, but would be spread laterally. As in the Airy system, the compensation is provided by the difference in density between crust and mantle. Vening Meinesz adopted a theoretical form for the downward deflection of the crust produced by a concentrated load, but left the distance at which the deflection became zero (radius of regionality) as a parameter to be determined. He prepared tables which allowed isostatic anomalies to be calculated for different crustal thicknesses and radii of regionality.

Gunn (1937, 1945), in a series of papers, developed the idea that the base of the crust be a true surface of equal pressure, with vertical stresses in the crust included. He termed this modification of the Airy mechanism "isobaric compensation", and indicated how isobaric anomalies could be computed. Gunn's work included a detailed analysis of the expected distortions of the crust under various conditions, so that the vertical crustal stresses could be calculated.

There are attractive features in both the Vening Meinesz and the Gunn systems. However, the above discussion on the importance of density variations below the Mohorovicic discontinuity applies to them also, and it is probably impossible at present to say that they are a closer approximation to the actual situation than the simple Airy mechanism. Jeffreys has noted that local Airy compensation makes fewer demands on the strength of the crust than do other models.

CHAPTER 7

Gravity Anomalies and Structures in the Earth's Crust

THE structures to be discussed in this chapter are chosen primarily for their importance in the study of the development of the earth's surface features. At the same time, the examples will indicate the approach that has been followed by various workers in the interpretation of anomalies. Some attention must first be given to the variation of density among rock types.

Rock Densities

We have so far indicated the variation of rock density with elastic wave velocity only. For the study of geological structures in the outer part of the earth it is necessary to have a correlation between density and rock type. The density of any rock depends partly on the densities of the minerals of which it is composed, and partly on the proportion of its volume which is occupied by pores. In the case of a very porous rock, the density will also vary depending on the extent to which the pores are filled by water.

The densities of common rock-forming minerals are well established, and typical values are:

Mineral	Density g/cm³
Quartz	2·65
Calcite	2·57
Orthoclase	2·57
Plagioclase	2·63 (Albite)–2·77 (Anorthite)
Biotite	3·0
Hornblende	3·2
Olivine	3·3
Serpentine	2·2

In the case of igneous rocks, the porosity is usually so low that the rock density is virtually the weighted mean of the mineral densities in the rock. For example, the density of a granodiorite with known mineral composition by volume may be computed as follows:

Mineral	Volume %	Density
Quartz	20·2	2·65
Plagioclase	47·6	2·67
Orthoclase	25·1	2·57
Biotite	7·1	3·00

Rock density 2·66 g/cm³

Acid igneous rocks generally have densities between 2·65 and 2·67 g/cm³. Density increases toward the basic end of the igneous rock scale because of decreasing quartz and orthoclase content, the increasing basicity of the plagioclase, and the increase in proportion of biotite, hornblende, pyroxene and olivine. The rocks of highest density found at the earth's surface are dunites, composed almost entirely of olivine. However, the minerals of basic and ultrabasic rocks are often altered in part to serpentine, a mineral of low density, and in these cases the rock density is much reduced.

To investigate the effect on density of porosity in a rock, let the fraction of the volume occupied by pores be P. Then, in a sample of mass m and volume v, the dry density ϱ_d is

$$\varrho_d = \frac{m}{v}. \tag{7.1}$$

But if the true density of the mineral grains (assumed constant) is ϱ_g, the volume of mineral present is m/ϱ_g. The volume of pores is

$$v - \frac{m}{\varrho_g}.$$

$$\therefore P = 1 - \frac{\varrho_d}{\varrho_g}$$

or
$$\varrho_d = \varrho_g(1 - P). \tag{7.2}$$

If the pores are partially filled with water, the density is increased. Let a be the fraction of pore volume occupied by water. Then the wet density ϱ_w is

$$\varrho_w = \varrho_d + P.a. \tag{7.3}$$

For example, in limestone the porosity may reach 30 per cent. The mineral calcite density is $2 \cdot 70$ g/cm³, but the dry density is $2 \cdot 7(1 - 0 \cdot 30) = 1 \cdot 90$ g/cm³. If the pores were completely water-filled, the wet density would be $2 \cdot 20$ g/cm³. There is thus a possible range in density from $1 \cdot 90$ to $2 \cdot 20$ g/cm³ in this one rock, depending on the fluid content.

Density is normally measured by weighing a sample first in air, then in water, and using Archimedes' principle to determine the volume. Precautions must be taken in the case of a porous sample to prevent water from entering the pores when it is immersed. The normal procedure is to weigh the sample in air, coat it with melted paraffin wax, weigh again in air, then immerse and weigh in water. The dry density is then

$$\varrho_d = \frac{w_a}{w_p - w_w - (w_p - w_a)/\varrho_p}, \tag{7.4}$$

where w_a, w_p, w_w are the weights in air, in air with paraffin wax, and in water respectively, and ϱ_p is the density of paraffin wax.

The mineral density may be determined by reducing the rock to a powder, and measuring the density of this. A more straight-forward procedure is to determine the water-saturated density as well as the dry density. The mineral density and porosity are then found from equations (7.3) and (7.4). In most cases the pores will not fill with water if the sample is simply placed in water to soak. But if the sample is placed in a chamber which can be evacuated, kept under vacuum for a few minutes, then covered with water before atmospheric pressure is restored, very nearly complete filling is achieved.

Porosity tends to decrease with increase in confining pressure, and therefore with the weight of overlying material on a rock. Sedimentary densities thus increase from very low values (2·0 g/cm³ or less) in the case of unconsolidated ocean-bottom deposits, or very near-surface material on land, to values approaching the mineral densities in formations near the base of a deep basin. Shale and clay in particular exhibit a remarkably

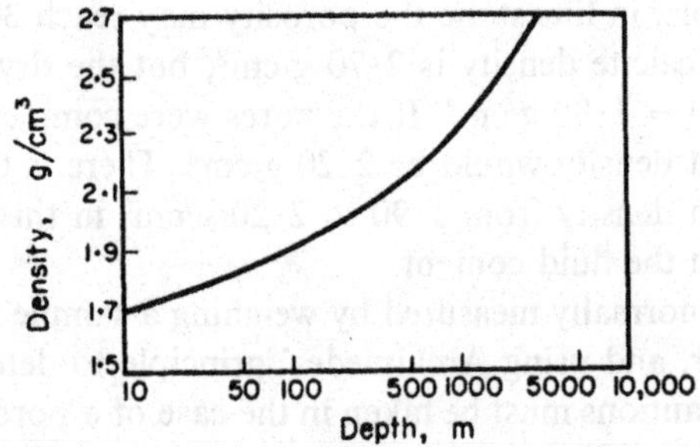

FIG. 7.1.   Variation in density of a clay with depth of burial (after Cook).

linear variation of density with depth of burial (Fig. 7.1), a fact which leads to interesting problems in the calculation of the attraction of a section consisting largely of shale (Cook, 1952).

Ocean Trenches

The most striking features discovered as a result of the first gravity measurements made at sea by Vening Meinesz (1929) were the narrow strips of intense negative anomalies observed over certain ocean depths. Later surveys showed that similar negative anomalies are found in many parts of the world, characteristically near the ocean margins. Figure 7.2 shows the gravity field in the vicinity of the Indonesian archipelago, and indicates the association of the anomaly with the ocean deep. The contours are based on isostatic anomalies, but the main negative effect is so

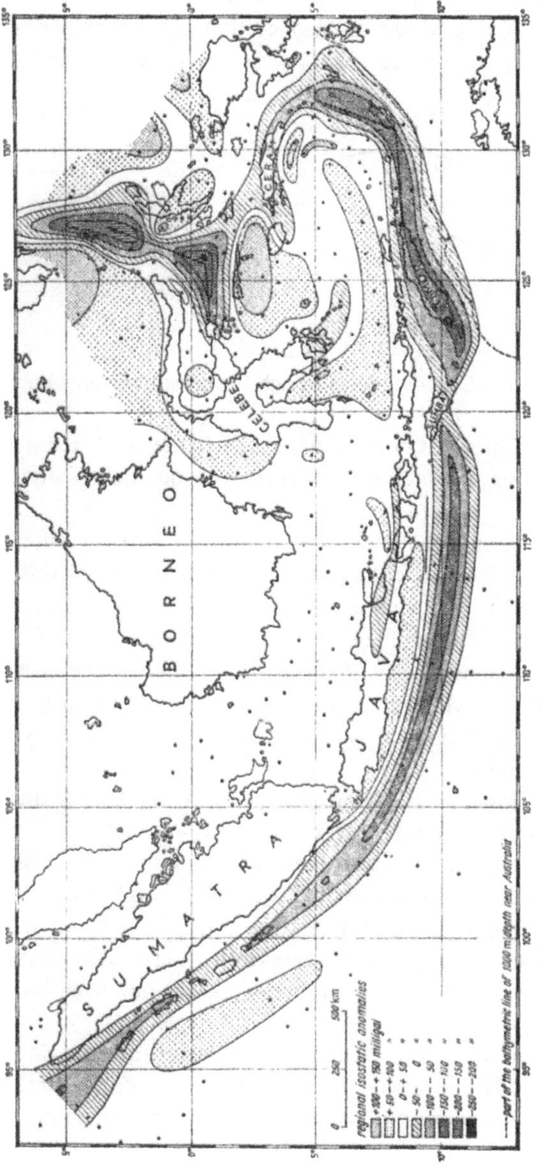

FIG. 7.2. Isostatic gravity anomalies in the East Indies (after Vening Meinesz).

intense (up to 200 mgal) that it is observed with any type of reduction. Vening Meinesz reasoned that the occurrence of these features near island arcs containing highly folded sedimentary rocks, and also near volcanoes, must be indicative of their

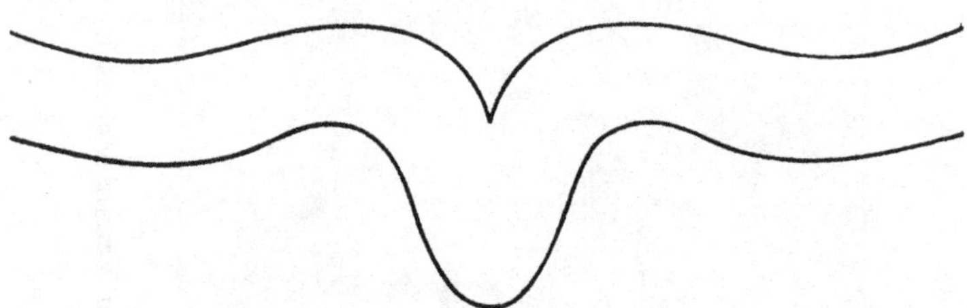

FIG. 7.3.   Cross-section through the crust after buckling.

relation to important horizontal stresses in the crust. He suggested that the uncompensated mass deficiency results from a downbuckling of the earth's crust, as the first stage in a cycle of mountain building. Taking the mass deficiency to be represented by a thickening or downbuckling of the crust, the form of this can be estimated from the shape of the gravity anomaly. If only the oceanic part of the profile is considered, and if a crust of constant density is assumed, a symmetrical thickening of the crust is inferred. Vening Meinesz considered that the thickening resulted from a combination of buckling and plastic flow toward the buckled region.

Crustal downbuckles (Fig. 7.3) revealed by gravity anomalies appeared to be definite evidence of the importance of compressive stress in the earth, and their significance to mountain building and geological processes in general was fully developed by Vening Meinesz himself (1954) and by Hess (1938). Recent seismic work at sea has shown that in some cases at least crustal buckling alone is not the explanation for the major part of the negative gravity effect.

Investigations of the Puerto Rico Trench (Talwani, Sutton and Worzel, 1959) and Tonga Trench (Talwani, Worzel and Ewing, 1961) have shown that density variations in the crust must be con-

sidered. In the case of the Puerto Rico structure, five velocity layers
were observed above the Mohorovicic discontinuity by seismic refraction, with compressional velocities 1·54, 2·1, 3·8, 5·6 and 7·0
km/sec. However, the seismological observations did not provide a
complete picture of the deformation of the base of the crust under
the Trench, and under Puerto Rico itself. Talwani, Sutton and
Worzel adopted densities for these five layers on the basis of the
velocity–density relationship, and adjusted the interfaces until the
calculated gravitational attraction agreed with the observed free-air
anomaly profile. Figure 7.4 indicates the gravity profile and the deduced section, with the control available from seismic refraction
also shown. Since the comparison is made with free-air anomalies,
the attraction of the water layer must also be included, and there

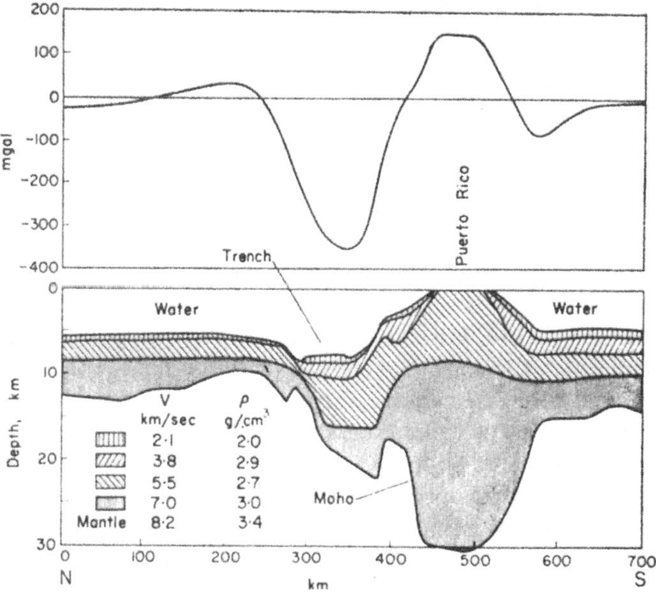

Fig. 7.4.   Crustal section through the Puerto Rico Trench determined
from seismic refraction and gravity profile (after Talwani,
Sutton and Worzel).

is a significant contribution to the negative anomaly from the accumulation of low-density material in the Trench. The result is an excellent example of the advantage of combining gravity and seismic measurements to produce the most complete picture of crustal conditions. Certain differences in general form with the interpretation shown in Fig. 7.3 for the Indonesian anomaly are apparent. On the seaward side of the Trench, the base of the crust is at a depth of 12 km or less. The crust certainly thickens under the Trench, but hardly in the form of an isoclinal fold or downbuckle. Compared to the oceanic side, the non-sedimentary crust under the Trench is about 4 km thicker, over a width of 100 km. The crust continues to thicken beneath the island of Puerto Rico itself. Similar results were obtained over the Tonga Trench, where the crustal deformation under the Trench itself appeared in the form of a flexure on the edge of the crustal thickening beneath the Tonga Ridge, rather than as a downbuckle. These results do not reduce the importance of trenches and associate island arcs in tectonic studies, but they do indicate that a different mechanism of formation should be considered. In particular, the section illustrated in Fig. 7.4 is suggestive of downfaulting beneath the Trench.

The case of gravity anomalies over rift valleys within the continents may be mentioned here. In a classic study of the gravity field over East Africa, using pendulum measurements, Bullard (1936) noted the occurrence of prominent negative Bouguer and isostatic anomalies over the great rift valleys of that continent. He concluded that these negative anomalies resulted from the downfaulting of strips of the crust between parallel thrust faults. This would imply that the rifts were formed by compressive forces in the crust. The distribution of stations available to Bullard was not sufficient to allow the field to be mapped continuously along the length of the structures. More recent work in Africa (Sutton, 1960), and also measurements over the Rocky Mountain Trench in Canada (Garland, Kanasewich and Thompson, 1961), have shown that the negative anomalies associated with rift-like structures tend to occur as separated, closed minima. As the greatest negative effects are observed where there are no bedrock exposures, it appears probable that a considerable part of the anomalies are due to low-density fill, in depressions of the rift floors, as in the case of the Puerto Rico Trench. It is still possible that an effect due to crustal distortion is present, but the determination of the magnitude of this will require a very careful correction for near-surface effects.

The Mid-Atlantic Ridge

The Mid-Atlantic Ridge is a broad swell passing down the middle of the Atlantic Ocean, from north-east of Iceland to the south Atlantic, where it apparently connects with similar ridges in the Pacific and Indian Oceans. Taken together, these mid-ocean ridges constitute the longest continuous geological structure on the earth, and they must be of great significance in the study of the development of continents and oceans. The central part of the Mid-Atlantic Ridge is marked by a rift valley, 20 to 30 km wide. Earthquake epicentres are concentrated along the central rift, and high values of heat flow from the earth's interior are measured along it. Heezen (1960) has pointed out that the central rift is very suggestive of formation by tension in the crust. The high heat flow is consistent with the hypothesis that the ridge is located over rising mantle convection currents. On this hypothesis, the Ridge could be thought of as formed of material brought up from the mantle along the tension crack formed by the spreading convection currents.

Information on the structure of the Ridge is of importance, and, as in the case of the ocean deeps, can best be obtained by a combination of seismological and gravimetric measurements. Talwani, Heezen and Worzel (1961) have analysed the gravity profile shown in Fig. 7.5, using as control the two refraction seismic profiles whose locations are shown. The free-air gravity profile across the central part of the Ridge is slightly positive, ranging from 0 to 50 mgal. It is very irregular, obviously reflecting the attraction of the irregular ocean bottom topography. The relatively small positive anomaly shows that the Ridge is almost

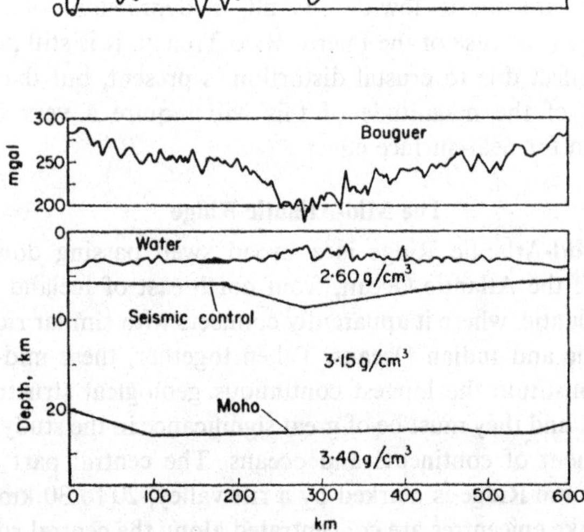

FIG. 7.5. Gravity anomaly profiles and deduced section across central portion of the Mid-Atlantic Ridge (after Talwani, Worzel and Heezen).

compensated, for if it consisted of an additional load on the deep (5 km) ocean floor it would produce a free-air anomaly of 250 mgal. This is borne out also by the depression in the Bouguer anomaly, which indicates a mass deficiency at moderate depths beneath the Ridge. The fact that certain narrow features, such as the central rift, produce much larger free-air than Bouguer anomalies shows that they are uncompensated locally, and that the short wavelength free-air effects are indeed the direct effect of the ocean bottom topography.

Talwani, Heezen and Worzel proceed to offer explanations for the mass deficiency, using the layers found by seismic refraction (Ewing and Ewing, 1959) as a starting point. The highest velocity observed in the seismic profiles was 7·30 km/sec, underlying a layer of velocity 5·00 km/sec, and a thin sedimentary cover of velocity 1·75 km/sec. Unfortunately, the base of the

$7 \cdot 30$ km/sec material was not observed, and there is no information on the manner in which it grades into the normal mantle. The above authors assume that the change to normal mantle, of density $3 \cdot 40$ g/cm³, is abrupt. The mass deficiency can then be explained by down-warping of the bases of the two crustal layers. No unique solution is possible, but a possible one, with its computed effect, is shown in Fig. 7.5. In this case, compensation is achieved partly by a depression of the Mohorovicic discontinuity and partly by a thickening of the $2 \cdot 60$ g/cm³ layer. The nature of the material of these layers is obviously important. It will be noted that neither layer has the typical oceanic velocity and density, as these are $7 \cdot 0$ km/sec and $2 \cdot 84$ g/cm³ respectively. In the upper layer both parameters are too small, and in the lower material both are too large. Ewing and Ewing (1959) suggest that the upper material is basaltic, presumably young basalt, not compacted to the extent of the normal oceanic crust. The lower material is considered to represent a mixture of basalt and mantle material. The Ridge does appear, therefore, to be built of volcanic material newly added to the ocean floor, the basaltic material presumably being derived by partial melting of the mantle at depth. The suggestion of a thick layer of mixed basalt and mantle material is interesting in view of the discussion in Chapter 6 on the nature of the mantle beneath mountain ranges.

Other solutions suggested by Talwani, Heezen and Worzel ascribe the mass deficiency to different warpings of the interfaces. The sharpness of the negative Bouguer anomaly demands that the major part of the mass deficiency lies within about 20 or 30 km of the surface. It does not appear possible to relate the entire anomaly to density variations produced by deep convection currents, although the broad decrease in background level, visible in Fig. 7.5, might be so explained. On the other hand, it is rather difficult to explain the apparent coincidence of down-warp and tension on the basis of convection currents, for (Chapter 6) tension and uplift should occur together. However, in the crust, the horizontal stresses due to convection currents may be dominant, and the down-warp could be a secondary feature.

Granite Batholiths in Continental Areas

The use of the term "granitic layer" for the upper part of the earth's crust suggests that granite itself should have properties typical of this layer. Also, we have noted that the density of granite, usually near $2 \cdot 67$ g/cm³, is the density often employed in making Bouguer reduction for material above sea level. Consequently, it was with a good deal of surprise that geophysicists first noticed large negative Bouguer anomalies over granites in many parts of the world (e.g. Cook and Murphy, 1952). There is by now no doubt that it is "normal" to observe a relatively negative effect over bodies of granite, and the only conclusion is that the upper part of the crust is significantly denser than true granite.

The problem was complicated, in the case of the Precambrian shields, which many consider to be typical of the continental crust beneath any sedimentary cover, by the former practice of mapping large areas of metamorphic rock as "granite". It is now evident that true, unfoliated granite is a rock of much more limited occurrence. The average density of the metamorphic rocks exposed in shield areas is probably of the order of $2 \cdot 76$–$2 \cdot 78$

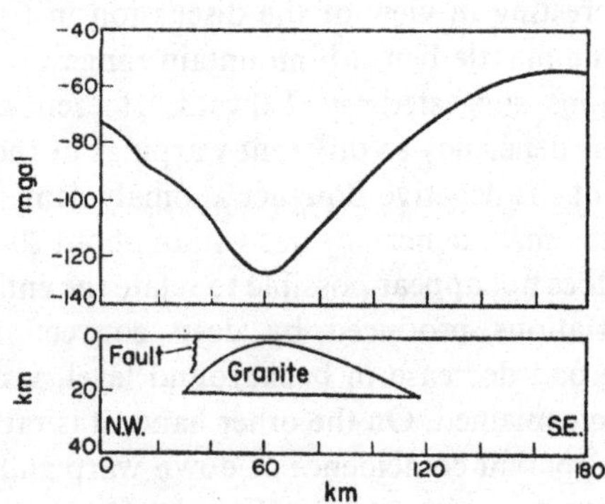

FIG. 7.6. Negative Bouguer anomaly observed over Canadian Precambrian Shield, presumably related to a granite mass (after Innes).

g/cm³, so that there is a density deficiency of 0·1 g/cm³ in the case of massive granite.

In the case of granite bodies exposed at the surface, the gravity anomaly can be used to estimate the extent in depth of the body. Innes (1957) has described a remarkably large negative Bouguer anomaly, exceeding — 120 mgal, from the eastern part of the Canadian Precambrian Shield. A profile across the feature is shown in Fig. 7.6. In plan, the negative anomaly has a length of approximately 200 miles. While exposures of granite at the surface are not definitely known on the line of section of Fig. 7.6, they are known to exist along the strike of the negative anomaly, some miles from this section. Innes assumed a density deficiency of 0·1 g/cm³, and adjusted the form of the granite mass so that its computed effect agreed with the observed profile. It will be noted that the mass is shown as spreading out with depth, and that the maximum thickness of granite is over 25 km. This represents a very large fraction of the total thickness of the crust beneath this part of the Shield. The granitic mass was emplaced south of, and parallel to, a major fault zone which marks the boundary between two provinces, of different ages, in the Precambrian.

The presence of large negative anomalies over bodies of granites has an important bearing on the origin of this rock. Two contrasting theories of the origin of granite batholiths are that they were intruded as granite magma produced by the melting of an original granitic layer, and that they were formed by the granitization of the pre-existing crustal rocks through the action of mineralizing solutions from below. The latter process would involve an upward concentration of the less dense minerals, and a settling of the denser constituents of the original rock. A theory based on the melting of a granitic layer must assume that the layer exists beneath the uppermost crust, since the latter is not typically granitic. Bott (1961) has suggested a mechanism by which granite bodies could be produced in this manner, and has pointed out that there is seismological evidence from some areas for a low-velocity layer within the crust.

There are certain difficulties to the assumption that the granite bodies were formed in place by differentiation. If a section of crust (Fig. 7.7) is differentiated to produce a low density region at the surface, there will certainly be a negative effect on gravity, as mass is removed from the surface. But if the body is of the same order of width as the crustal thickness, the attraction of the denser fraction cannot be appreciably diminished, and there is a limit to the negative anomaly that could be produced, unless crustal thickening occurs in conjunction with the granitization. As negative anomalies greater than this maximum are observed, for bodies of comparable size to that shown in Fig. 7.7, it must be assumed that the denser fraction is somehow separated from the resulting granite, other than by settling. The granite case thus provides an excellent example of the direct application of gravity measurements to geological problems. It also emphasizes the need of the geophysicist to have truly representative mean densities for large areas of the continental crust.

Sedimentary Basins

We have mentioned the accumulation of sedimentary rocks in oceanic troughs, and the use of gravity measurements to provide

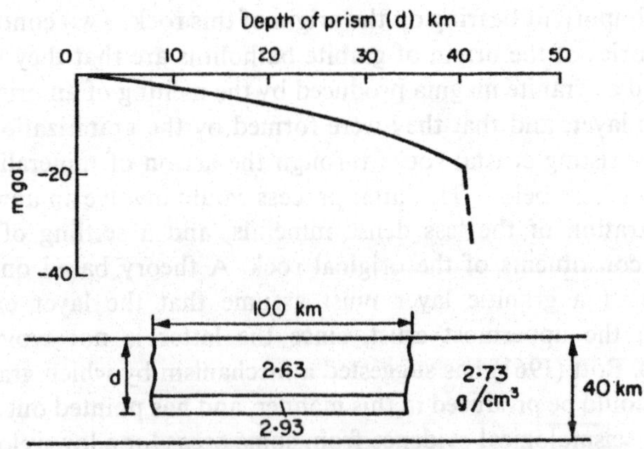

FIG. 7.7.   Gravity anomaly that could be produced over a differentiated section of crust.

information on crustal conditions associated with the depressions. The structure of sedimentary basins which now occur within continental areas can also be investigated effectively by geophysical methods.

It has been realized for some time that an active mechanism within the earth must have been operative at the time of deposition of great thicknesses of sedimentary rock. Since it can be shown by the methods of sedimentology that, in many instances, entire sections were deposited in shallow water, the sediments in these cases could not be assumed to have been dumped into pre-existing basins. Early ideas of isostasy then suggested that the crust would be depressed by the load of sedimentary material added to the ocean bottom, and for a time this depression by weight was an accepted explanation. However, it is not difficult to show that a maximum thickness of accumulation is soon reached under these conditions. Let us consider a broad oceanic area, one unit deep, initially in isostatic equilibrium. We assume that sedimentary material, of density $2 \cdot 5$ g/cm^3, is deposited in the ocean until it is filled, and also make the assumption that the crust is depressed a distance x, so that outflow of mantle rock, of density $3 \cdot 3$ g/cm^3, provides compensation. Then, equating the mass added to the mass of water and mantle displaced, we have

$$(1 \mid x)2 \cdot 5 - 1 \cdot 1 + 3 \cdot 3x$$

or
$$x = 1 \cdot 6. \tag{7.5}$$

The total thickness of section that can be accumulated is thus $2 \cdot 6$ times the original water depth. It would appear impossible for sections as thick as 10 km (which are found) to be deposited in shallow seas. In the case of narrow regions of deposition, the strength of the crust would have to be considered, and this would tend to reduce the maximum thickness. The only conclusion is that sedimentary rocks accumulate in regions of the earth which are pulled down from within.

Studies of the structure of sedimentary basins are useful in suggesting the mechanism by which this depression of the crust was accomplished. Since sedimentary rocks are usually less dense than the igneous and metamorphic rocks which form the floor beneath the sediments, conspicuous negative gravity anomalies are normally observed over basins. Over the central part of a broad basin, the gravity anomaly due to the basin itself should be given approximately by the Bouguer formula

$$\Delta g = 2\pi G h \Delta \varrho, \tag{7.6}$$

where $\Delta \varrho$ is the deficiency in density of the sedimentary rocks. For a difference in density of 1 g/cm^3, the thickness h required to produce an anomaly of 1 mgal is 24 m. An example of a basin structure is the Central Valley of Chile, over which the gravity field was investigated by Lomnitz (1959). A profile across the basin is shown in Fig. 7.8. Because conditions are quite uniform in the direction perpendicular to the profile, a two-dimensional approach was satisfactory. The rocks exposed to the west of the valley, and also eastward toward the Andes mountains, were found to have a density of 2·67 g/cm^3, while the sedimentary rocks (at least at the surface) had a density of 2·0 g/cm^3. The Bouguer anomaly profile decreases from west to east, but not all of this decline is due to the basin, as the eastern end of the profile is underlain by the denser rock. There is obviously a broader

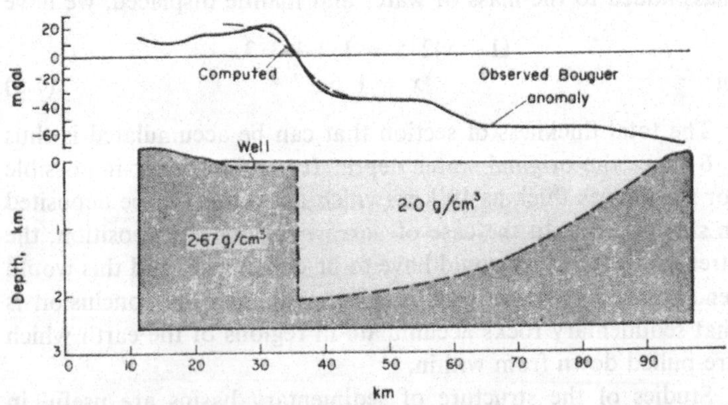

Fig. 7.8. Bouguer anomaly profile and deduced structure across the Central Valley of Chile (after Lomnitz).

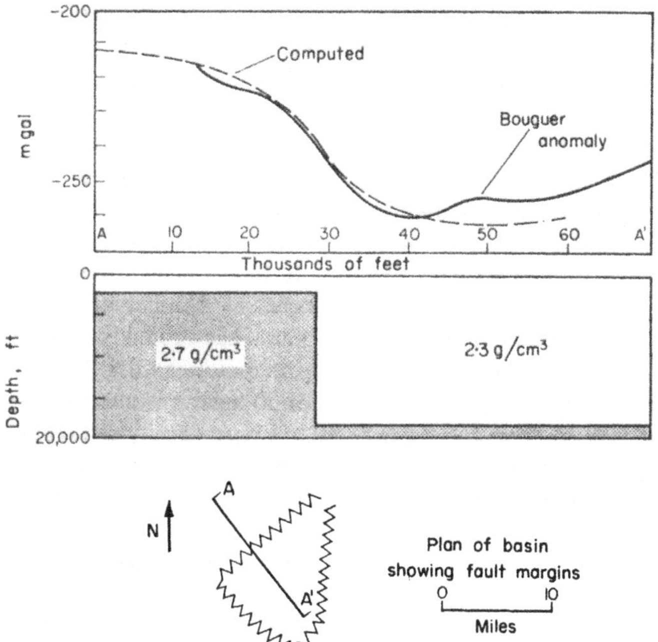

Fig. 7.9.   Observed and computed Bouguer anomaly profiles across the faulted margin of the Mono Basin, California (after Pakiser, Press and Kane).

influence present, probably the mass deficiency of the Andes' roots. However, the abrupt change in anomaly near the centre of the profile is proof of a near-surface structure, and Lomnitz interpreted this as the effect of a near-vertical fault. The thickness of sedimentary rock to the east of the fault is estimated to be 2·0 km, and the theoretical effect on gravity for a fault of this throw is shown in Fig. 7.8. This interpretation provides a good example of the possibility of reducing the ambiguity of the gravitational field, when some parameters are known. The thickness of sedimentary rock west of the fault was known from a single well, and the maximum thickness to the east could be estimated from the change in anomaly, using the Bouguer formula for a slab. Computations for faults of different dip then showed that the fault had to be nearly vertical to explain the steep gravity gradient.

Lomnitz pointed out that the absence of a similar abrupt change in anomaly level over the eastern part of the basin indicated a more gradual thinning of sedimentary rock in that direction. The deduced structure is thus very suggestive of a hinge or rotary movement of the crust about an axis east of the basin. Presumably the same forces in the crust responsible for the uplift of the Andes were responsible for the depression of the basement floor.

An example of a remarkably deep basin of limited areal extent is the Mono Basin of California. In plan it has a roughly triangular form (Fig. 7.9), with the greatest dimension hardly 20 km. A negative Bouguer gravity anomaly of 50 mgal was observed over it by Pakiser, Press and Kane (1960). Using a density deficiency of $0 \cdot 4$ g/cm^3 for the Cenozoic deposits in the basin, as compared to the surrounding older rocks, they inferred the faulted nature of the margins, shown also in Fig. 7.9. In this case the computed gravity anomaly curve over the edge was not based on a two-dimensional structure, but the basin effect was computed by dividing the sedimentary section into a number of horizontal slabs, and determining the solid angle subtended by each at points on the profile. Seismological and aeromagnetic measurements confirmed the general picture of a basin produced by downfaulting along three intersecting faults, and the authors concluded that subsidence was the result of the removal of material at depth, by vulcanism.

There is one difficulty often encountered in the use of gravity measurements to estimate the thickness of a sedimentary section. That is the determination of the true difference between the mean density of the sedimentary formations and the underlying rocks. Very often only the surface density of the sediments can be measured directly. Because of the effects of compaction (cf. section on Rock Densities), the density must increase with depth in the sedimentary section. The rate of increase will, however, depend on the lithology, being much greater for shales than for sandstones and limestones. It may be impossible, therefore, to estimate a representative density for the section as a whole. If the effects of compaction are not allowed for, of course, the basin depth will be underestimated.

Masses of Basic and Ultrabasic Rock

It has been suggested that the mantle of the earth, beneath the crust, may consist of rock consisting largely of olivine. Masses of rock of similar composition occur at the earth's surface, particularly in highly folded areas, and there is considerable interest in determining the form in depth of these masses. It is important to know if any of the masses extend through the crust to the mantle; if they do not, the mechanism by which mantle material is intruded through the crust raises problems. As the density of these basic, olivine-bearing rocks is much greater than that of the normal crust, conspicuous positive gravity anomalies are usually observed over them, and the gravity field can be used to provide information on the depth extent.

Extremely large positive anomalies have been known to exist over Cyprus for some years (Mace, 1939), and these have been examined in detail by Harrison (1956). The geology of the island is rather complicated, as there are basic rocks of different ages. Most significance is attached to intrusive bodies of olivine-bearing gabbro, partly altered to serpentine, which was emplaced into older diabase. The exposures of the gabbro are of moderate size, the largest known body having a length of 10 miles, but two exposures near the centre of the island are believed to be connected at depth. Relatively low-density sedimentary formations conceal the igneous rocks around the outside of the island.

Figure 7.10 shows the variation in Bouguer anomaly along a north-east–south-west line across Cyprus. There is relief of approximately 120 mgal from the coast lines to the peak of the anomaly. The positive effect is elongated, crossing the island, from north-west to south-east, and apparently trending out to sea. Harrison used a two-dimensional approach to investigate the form of high-density material responsible for the gravity anomaly. The interpretation is made more difficult by the presence of the low-density sedimentary rocks on the flanks of the island, as these must diminish the gravity field. However, the body shown in Fig. 7.10 is typical of the general form found by Harrison. It consists of a relatively narrow section extending well into the crust, and a shallow, flatter portion. This form is what would be expected if the ultrabasic material were pushed up along a central feeder, and then spread out laterally as it reached the surface.

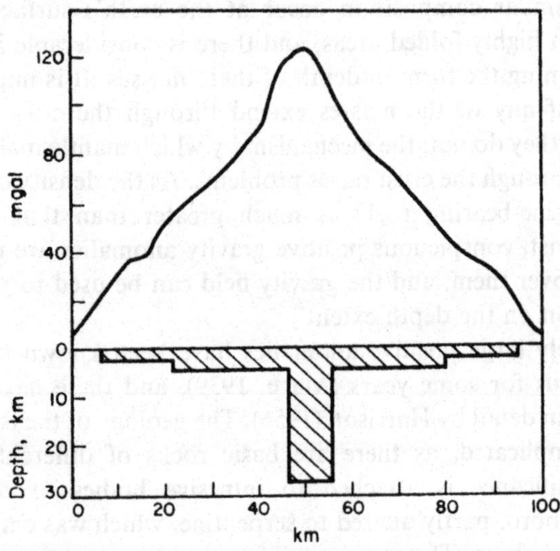

FIG. 7.10. Bouguer anomaly profile across Cyprus, and inferred distribution of ultrabasic rock (after Harrison).

There remain uncertainties as to the mechanism by which the dense material was forced upward through the crust. The point here is that gravity measurements over ultrabasic rocks provide information on the concealed parts of the masses, and any theory of intrusion must explain the form which is found. There is a second significance to the positive anomalies found over Cyprus. We have mentioned above the Bouguer anomalies, but over the land the isostatic anomalies are at least as large. The area is thus one of great mass excess locally, the ultrabasic material representing an uncompensated load on the crust. The island might well be assumed to be sinking on this account. However, it has not sunk in historical time, and the thick sections of Tertiary sedimentary rocks now above sea level around its margins show that it has risen since Tertiary time. This indicates that areas of positive gravity anomalies cannot be assumed to be sinking. The mass excess per unit area over Cyprus is considerable, but the region is sufficiently narrow for stress systems in the near-surface rocks to support the load.

Areas of Post-glacial Uplift

It is well known that there are some areas of the earth over which the elevation of the land surface is rapidly changing. Repeated levelling in Finland has shown that, along the Gulf of Bothnia, the uplift is proceeding at the rate of 9 mm per year. Since this area is known to have been near a centre of Pleistocene glaciation, it has been assumed that the present uplift is a recovery from depression produced by ice loads on the crust. If this is the case, the rate of recovery, which is related to the rate of flow of mantle material, can be used to estimate the viscosity of the mantle (Niskanen, 1939; Haskell, 1935). If the material of the mantle can be treated as a viscous fluid, the viscosity is a parameter of great importance in the study of earth processes. The values of viscosity obtained from considerations of the Fenno-Scandian uplift were 1×10^{22} poises by Haskell, and $3 \cdot 6 \times 10^{22}$ poises by Niskanen. It is necessary for these analyses to know not only the present rate of rise, but also the amount of recovery still required, since the mechanism, if viscosity is dominant, would be one in which the rate varied with time. The variation in height with time should follow a relaxation curve. Gravity measurements have been used to infer the amount of uplift still required for compensation.

Let us consider, as in Fig. 7.11, a section of crust initially standing a height h above sea level. On the basis of Airy isostasy, the crustal thickness is

$$T + h \cdot \frac{\varrho_n}{\varrho_n - \varrho},\tag{7.7}$$

E

where T is the crustal thickness beneath an area at sea level, and ϱ_n and ϱ are the densities of the mantle and crust respectively.

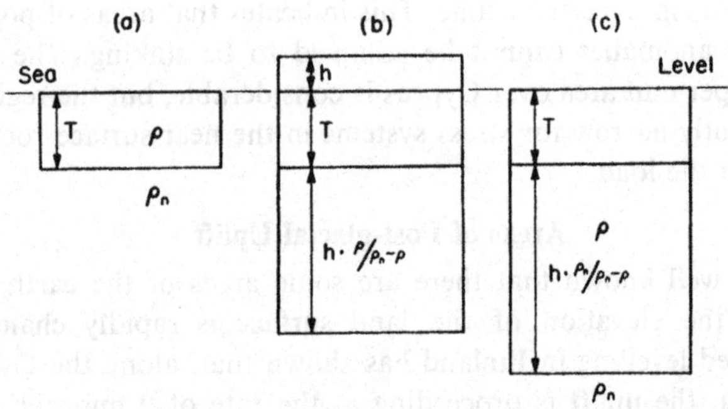

FIG. 7.11. Sections of (a) normal crust at sea level, (b) crust beneath an elevated area, and (c) crust beneath a depressed region.

Suppose the region is depressed by the weight of ice until the land surface is at sea level. After the ice has melted, the area has an anomalously thick crust for a region at sea level, the mass deficiency per unit area being

$$h \cdot \frac{\varrho_n}{\varrho_n - \varrho} \cdot (\varrho_n - \varrho) = h \cdot \varrho_n. \tag{7.8}$$

But $h \cdot \varrho_n$ is simply the mass per unit area of mantle material which was pushed aside during the depression of the crust. Over a broad area, a negative isostatic gravity anomaly would be observed, of magnitude

$$\Delta g = 2\pi G \varrho_n h. \tag{7.9}$$

It would appear, therefore, that if Δg could be determined, the height h of uplift required to restore compensation would be known. The difficulty is that the anomaly field over a broad area (which is required) contains contributions from many sources, and the estimate of that portion due to crustal depression is by no

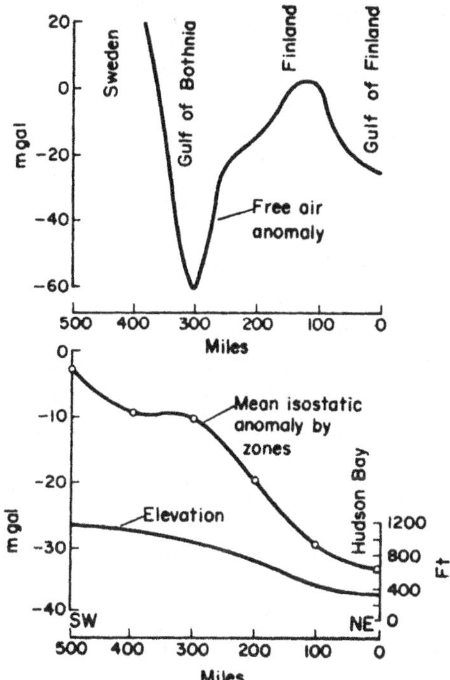

FIG. 7.12. Gravity anomaly profiles from Fenno-Scandia (top, after Honkasalo) and Canada (after Innes), showing negative trends which have been related to glacial depressions.

means easy. Figure 7.12 shows profiles from two areas, in which some of the effect does appear to be related to this cause. One profile shows the variation in isostatic anomalies across Finland, toward the Gulf of Bothnia, where the present rate of uplift is a maximum. The negative anomalies reach an amplitude of 50 mgal, and from these Niskanen estimated that the land had 200 m to rise before compensation was restored. The simplified equation (7.9) above would give the same result if the mean negative anomaly is taken to be 28 mgal. However, as Honkasalo (1959) has pointed out, the gravity field of Fenno-Scandia consists of a series of positive and negative trends, which are elongated parallel to the structural features of the region, but are not concentric around the centre of uplift. A single profile, such as that plotted in Fig. 7.12, can therefore be misleading. The second profile shown in Fig. 7.12 is from the Hudson Bay area of northern Canada (Innes, 1960). Innes determined the mean anomaly over concentric zones, 100 miles wide, around the Bay, and as each of these zones cuts across many structures, the means are probably relatively free of the effect of these. (Some features in the Pre-cambrian shield, however, produce systematic effects on gravity which do maintain one sign over several hundred miles.) The decrease in mean isostatic anomaly toward Hudson Bay, from values which are close to zero at a distance from it, is striking. On the basis of equation (7.9), the negative effect is equivalent to a depression of about 230 m.

It should not be assumed that uplift will actually continue until compensation is restored. The centres of depression in both Fenno-Scandia and Canada are probably sufficiently restricted in extent for a moderate deficiency in mass to be maintained by near-surface stresses. Conversely, there are almost certainly changes of level now in progress which are not related to surface loads, or the removal of loads. These are presumed to be brought about by active force systems in the crust or mantle.

Existing Glaciers

Glacial ice has a density of approximately $0 \cdot 90$ g/cm^3, and a valley filled with ice therefore has a mass deficiency in comparison with crustal sections on either side. Gravity measurements, in conjunction with other geophysical methods, have proved useful in determining the form of the bedrock floor beneath glaciers. These studies are of value in the estimation of the total volume of ice presently contained in a given glacier, and in providing back-ground information necessary for studies of glacial flow.

The problem is complicated by the fact that a gravity profile must extend on to the solid rock on either side, in order that the true change in anomaly due to the ice is determined. In the case

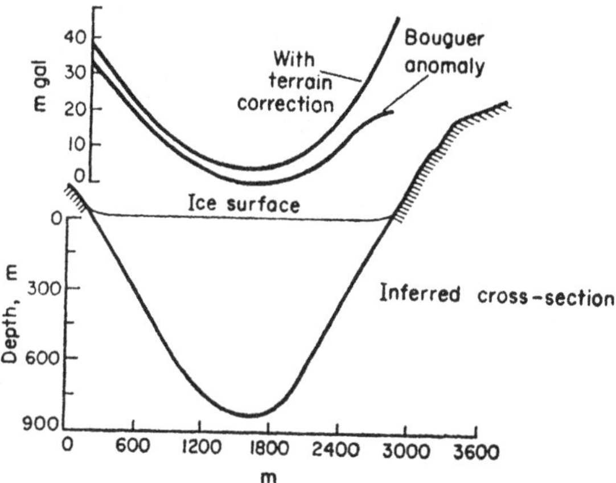

FIG. 7.13. Bouguer anomaly profile and deduced cross-section for the Salmon glacier, British Columbia (after Russell, Jacobs and Grant).

of valley glaciers, the sides may be of difficult access. Even if stations can be established on rock at the edges, the terrain effects at these stations may be so great that an accurate computation is impossible.

An example of a profile across the Salmon Glacier (Russell, Jacobs and Grant, 1960) of north-western Canada is shown in Fig. 7.13. The approach used here was to compute Bouguer anomalies for all stations, using a density of $2 \cdot 67$ g/cm^3. The ice was then considered to be an anomalous body, with density deficiency $1 \cdot 77$ g/cm^3. As the valley was relatively uniform in width along its axis, two-dimensional tabular models were used both for the terrain corrections and for the interpretation. The illustration shows the very large terrain corrections which were required at either side. The cross-section was finally deduced by a process equivalent to assuming it to be composed of a number of slabs with vertical edges. A graticule, such as described in Chapter 5, could also have been used to determine the shape which gave the best fit.

Over extensive glaciers or ice-caps, gravimetric methods for thickness determination are most effective when some control is available, in the form of holes drilled through the ice, or seismic profiles. Because of the rapidity with which gravity measurements can be made, and the portability of the equipment, there is a considerable advantage in reducing the drilling and seismic work required on a glaciological expedition.

Meteor Craters

The final type of crustal feature over which gravity measurements will be considered is the meteorite crater, or suspected crater. It has been suggested (Beals, Ferguson and Landau, 1956) that there are analogies between certain features of the earth and moon topographies. If this is the case, it might be expected that a number of circular depressions on the earth were formed by meteor impact. Since depressions on the earth's surface tend to become filled with sedimentary material, geophysical methods are useful in determining the true depth of these features. This depth is required in order to compare the profile of the feature with that of known craters, and also to estimate the meteorite energy which would be required to produce it.

An example of a gravity map over a suspected crater near Brent, Ontario, Canada, is shown in Fig. 7.14 (Innes, 1961). The negative anomaly of 5 mgal, is evidence of a circular body of low density beneath the topographic depression in the centre of the map-area. This type of anomaly lends itself to interpretation in terms of vertical cylinders for points on the axis, or by means of a solid angle chart for other points. The determination of the actual depth of the circular feature requires a knowledge of the density of the material filling it. Innes, in the case of this and other features, found that the mass deficiency results from both low density sedimentary fill, and also from a decrease in density of the original crustal rock due to brecciation. The latter apparently resulted from meteorite impact. In the case of a crater formed in Precambrian gneiss of density $2 \cdot 67$ g/cm^3, the brecciated material was found to have a density of $2 \cdot 50$ g/cm^3, which was very close

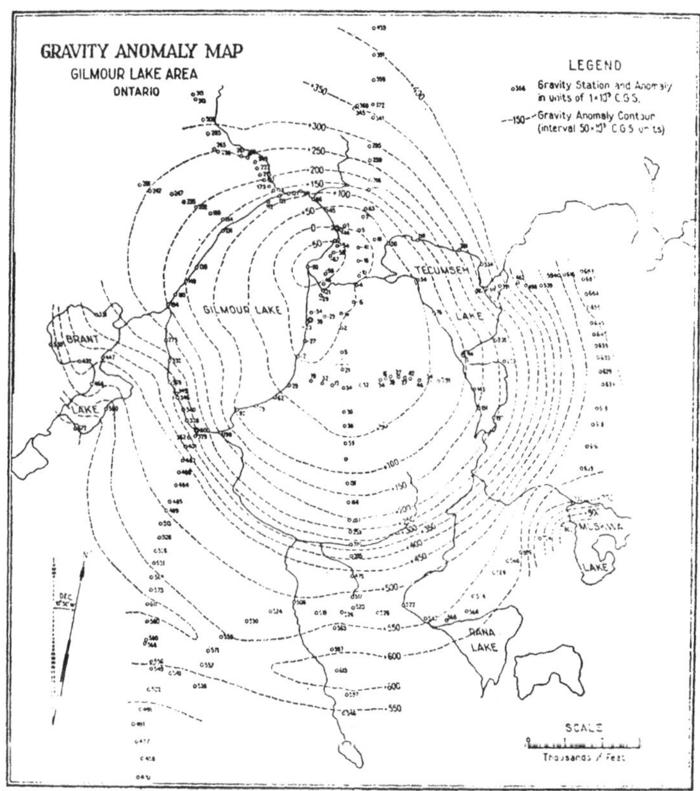

FIG. 7.14. Gravity contours over a circular feature in Canada which is known as the Brent crater. (After Innes. Illustration courtesy of Dominion Observatory, Ottawa.)

to the density of sedimentary fill. The Brent feature, including fill and the underlying brecciated zone, was estimated to extend to a depth of 5,000 ft. In the same paper, Innes described the investigation, also with gravity measurements, of a still larger feature at Deep Bay, Saskatchewan, Canada. The gravity survey, conducted on the ice of the Bay in winter, suggested a crater depth of 12,000 ft.

CHAPTER 8

The Place of Gravity Measurements in Geophysical Prospecting

THE USE of gravity surveys in the search for oil and minerals has had a most important influence on the development of gravimetry as a branch of geophysics. Much of the progress in the improvement of instruments was due to this incentive, and many of the techniques of interpretation that have been mentioned were first applied in prospecting. It is very desirable that close co-operation be maintained between those geophysicists engaged in geophysical prospecting and those who apply gravity measurements to other problems. Very often, a number of the gravity determinations made in the course of a survey could be released for academic purposes without risk of disclosing the local conditions of economic importance.

The general approach in prospecting is identical to the one we have already described for the study of broader structures of the crust. We will discuss here only the modifications which are required because of problems peculiar to applied geophysics. It should be borne in mind that the usual approach in prospecting is to use a combination of geophysical methods, together with geological examination and drilling. In petroleum exploration, gravity surveys are normally carried out in advance of detailed seismic work, while in mining exploration gravity measurements are usually made after preliminary investigation by electrical or magnetic methods. It is our purpose here only to point out the particular requirements of prospecting applications as far as gravity operations are concerned, and not to evaluate different geophysical methods.

Observations and Reductions

In most cases the structures sought in geophysical prospecting are of very limited extent, and produce anomalies of very small amplitude, compared to the crustal studies discussed in the previous chapter. The aim is thus to obtain a dense network of stations over a local area, with the highest possible relative accuracy in the Bouguer anomalies. Observations are normally made with a portable gravimeter, and all precautions are taken to reduce the errors due to drift. As far as the immediate purpose is concerned, it is not even necessary to tie the observations to a point where g is known absolutely. However, the value of the measurements for other applications is greatly increased if the determinations are on an absolute datum, and prospecting companies should be urged to do this.

The Bouguer method of reduction is used, but since relative values of anomaly are of prime interest, the method of reduction is usually not that used for broad-scale surveys. Corrections for station height are made by application of the free-air and Bouguer terms, but the correction is normally made to a datum elevation rather than to sea level. The datum elevation is chosen to be near the mean elevation of the stations, in order to simplify the numerical terms. Similarly, the variation of gravity with latitude is removed, not by direct comparison with the international formula, but by the use of a latitude variation term derived from it. If the derivative of the formula with respect to latitude is taken, and expressed as a space gradient on the earth's surface, the gradient obtained is $1 \cdot 307 \sin 2\varphi$ mgal/mile, where φ is the latitude.

The latitude correction in the case of local surveys is usually made by choosing a base latitude for the area, and correcting all stations on the basis of their distances north or south of this line.

The density used in making the Bouguer reduction is of considerable importance. Since the aim is to obtain a distribution of anomalies free from the effect of local topography, the density should be chosen so as to be representative of the local near-surface material. The value used may be based on sample measurements, although Nettleton (1939) has suggested the use of special gravity profiles across isolated topographic features of the area. If Bouguer reductions are then made with a number of trial densities, the density which gives the smallest correlation with topography can be chosen. Terrain corrections acquire an added significance because of the small amplitude of the anomalies sought, and even in non-mountainous areas the correction must be made with care. In any particular area, the accuracy with which the terrain effect can be computed should be estimated, and it should be established that the effect on gravity of the structure of interest is greater than this amount.

Isolation of Anomalies

The anomalies which occur in many prospecting applications appear as very minor flexures on the anomaly field due to broader crustal structures. It is impossible to proceed with interpretation until the local effect can be isolated from these broader trends. If the area of the survey is limited, the regional trend may appear as a uniform increase, represented by parallel, evenly-spaced contours (Fig. 8.1). A local anomaly, which ordinarily would be indicated by closed contours, appears as a "nose" on the regional field. Various methods have been used to effect the separation of the local effect, and two of these are illustrated in Fig. 8.1. These consist of manual smoothing, either on the contour map or on profiles, to estimate the undistorted regional field. At each station, the value of the smoothed field can be subtracted from the original Bouguer anomaly to give the residual anomaly. The disadvantage of these methods, obviously, is that they depend very much on the personal bias of the geophysicist, and small differences in drawing the regional field produce large relative differences in the residual anomaly. This bias is partially removed if a more mechanical method is used to estimate the regional field. If the latter can be thought of as a sloping "plane", the value at any point is equal to the average value of the anomaly around any circle, large enough to exclude local effects, drawn around the point. This suggests the use of a template, on which a number of points on such a circle

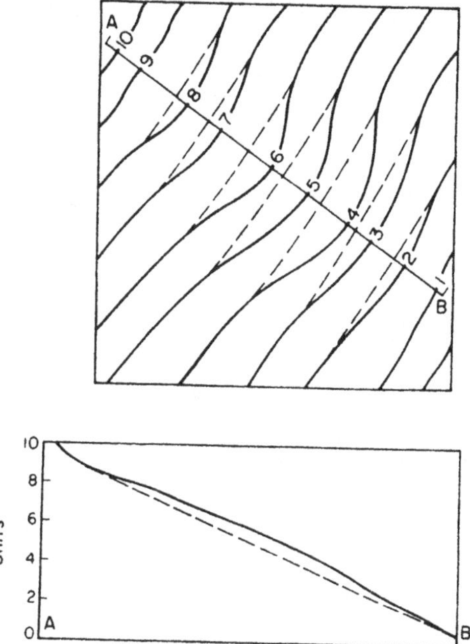

FIG. 8.1.   Separation of local and regional anomalies on the contour
map (top) and profile.

are marked, to obtain the circle-average around all stations of
the survey. It is true that the method is non-objective once the
circle diameter is selected, but it is found that different residual
maps are obtained when circles of different diameter are used. If
the diameter is too small, the circle-average does not remove all of
the local anomaly from the estimate of the regional field at the
centre; if too large, the assumption that the regional trend is
uniform may not be valid. The circle may be thought of as a
band-pass filter, fairly effective at isolating anomalies whose width
is of the same order as its diameter.

A quite different approach to the problem of isolating local
effects is the calculation of second vertical derivatives of the
gravitational field. Since the anomalous gravity, Δg, satisfies

Laplace's equation (Appendix 1), the second vertical derivative is related to the two horizontal second derivatives:

$$\frac{\partial^2(\Delta g)}{\partial z^2} = -\left(\frac{\partial^2(\Delta g)}{\partial x^2} + \frac{\partial^2(\Delta g)}{\partial y^2}\right). \tag{8.1}$$

Here, z is vertically downward, and x and y are two axes in the horizontal plane. Now the value at any point of $\partial^2(\Delta g)/\partial x^2$ is, very nearly, a measure of the curvature at that point of the anomaly profile taken parallel to the x-axis. The expression on the right of equation (8.1) is thus the total curvature, in two perpendicular directions, of the anomaly surface at the point. Normally, curvature is positive when a curve or surface is concave upward. The negative sign indicates that the second vertical derivative is positive when the anomaly surface is concave downward; that is, where there is a local maximum.

It is apparent that the curvature of the anomaly surfaces is much greater for local than for regional effects. If the regional trend across an area is represented by a plane surface, its curvature is zero. The second vertical derivative is therefore a powerful tool for isolating effects which might otherwise be unnoticed. Values of $\partial^2(\Delta g)/\partial z^2$ can be calculated by the methods outlined in Chapter 5, or by equivalent methods (Elkins, 1951), plotted, and contoured. A comparison between a Bouguer anomaly map and a second vertical derivative map is shown in Fig. 8.2, which indicates the striking accentuation of local anomalies. In the case of the anomaly over an isolated structure, the edges of the structure are often indicated on the gravity profile by the inflection points of the curve. These are points of zero curvature, consequently the zero-contour of the second-derivative map may be expected to outline the edges of local structures.

The calculation of second vertical derivatives is not without certain disadvantages, and it does not solve all problems connected with local surveys. It should be noted that the resulting quantities are no longer in gravity units. The dimensions are cm $^{-1}$ sec $^{-2}$, and, of course, profiles of the gravitational effect of

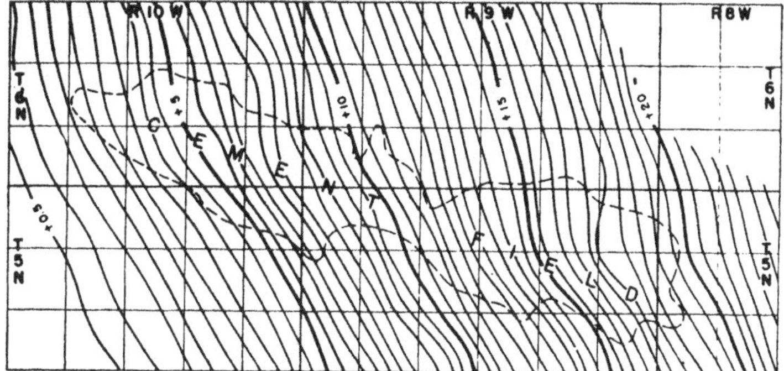

Cement field, Oklahoma. Observed gravity.
Contour interval, 0.5 milligal.

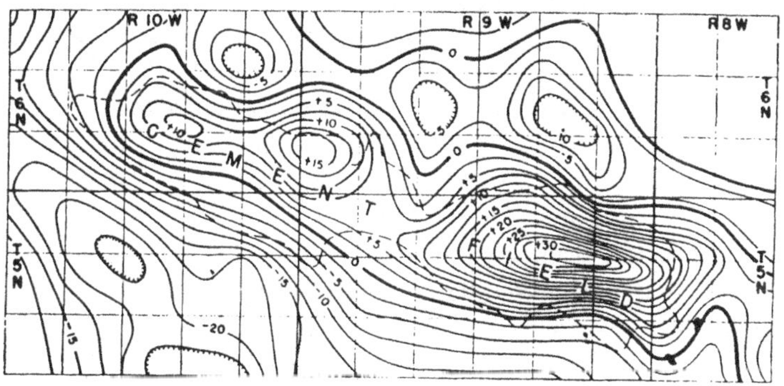

Cement field, Oklahoma. Second derivative. Contour
interval, 2.5×10^{-16} c.g.s.u.

FIG. 8.2.   Bouguer anomaly (top) and second derivation map of the
Cement field, Oklahoma. (After Elkins. Illustration courtesy
of Gulf Research and Development Company, and the
editor of *Geophysics*.)

model structures cannot be used in the interpretation of the
second-derivative map. However, profiles of the second-derivative
effect itself can be computed for simple shapes, and these can be
helpful for interpretation. A serious disadvantage is the fact that
random errors are accentuated in the process. In a dense network
of stations an effect which is random from station to station

produces very large changes in curvature. On the other hand, if the stations are not well distributed over the area, it is not possible to carry out the necessary surface integration with precision. The calculation of second vertical derivatives should be restricted to cases where there is an adequate distribution of stations, with anomaly values of high accuracy.

Exploration for Petroleum

The greatest application of gravity methods in geophysical prospecting has been in the search for oil and gas. These fluids occur in sedimentary basins, and tend to accumulate, in porous formations, where they are trapped by structural deformation of the formations (for example, in anticlines or domes), or by changes in porosity. Gravity surveys are of value in suggesting the location of structures favourable to accumulation.

In order of decreasing scale of problem, we may consider in turn the search for basins, the study of conditions in the crust beneath the sedimentary section, and the search for structures within the sedimentary rocks. We have already considered the gravity effect of sedimentary basins themselves in the previous chapter. It is only necessary to point out that in parts of the world which have not been geologically mapped in detail, the verification of the presence of a thick section of sedimentary rock is the first stage in reconnaissance exploration for petroleum. A broad-scale gravity survey will often indicate whether or not such a section is present.

Over a sedimentary basin, the larger gravity effects of fairly limited extent are almost invariably due to conditions within the crust beneath the sedimentary rocks. Normally, there are simply not the density contrasts within the basin to produce anomalies greater than a very few milligals. Moderate anomalies may be produced by topography on the basement surface, and these will be of interest if they are associated with conformable flexures in the overlying formations, which may represent possible petroleum traps. In the case of larger anomalies, it can usually be shown that no reasonable topography on the basement surface could produce them, and that they must be the result of density variations within the basement. The fact that this type of variation in density can extend to great depth in the crust permits relatively large anomalies to be produced. In this case, the only application to petroleum exploration is the estimation of the depth to basement beneath the area of the anomaly. For example, Fig. 5.12 showed the variation in form of the anomaly profile over a contact with the depth of the structure. Comparisons of observed profiles with computed effects such as these can provide a first estimate of depth to basement.

In the case of structures within the sedimentary section, which are of greatest interest, most success to date has probably been achieved where salt has been involved, either as domes or beds. Salt, with a density of $2 \cdot 2$ g/cm^3, is normally considerably less

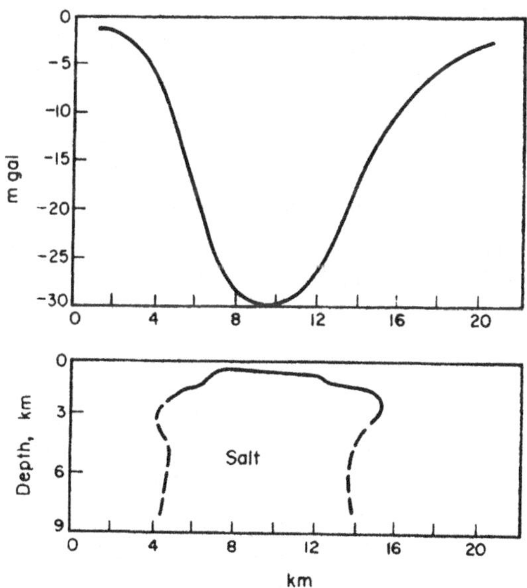

FIG. 8.3.   Bouguer anomaly profile across a salt dome. The top of the dome was known from seismic measurements; the remaining form was determined from the gravity profile.

dense than other sedimentary rocks, and the density contrast produces measurable anomalies. Figure 8.3 shows the gravity profile across a salt dome in Russia, with a comparison between the form as inferred from this profile and that deduced by seismic measurements and drilling. Salt domes, which have the approximate form of vertical cylinders, can produce circular gravity minima which may be analysed in terms of the form of the dome. Oil and gas may accumulate above the salt, or around the margins, where sedimentary formations were interrupted by the intruding mass.

Other structures in the sedimentary section generally produce much smaller effects. However, Yungul (1961) has shown that, in favourable cases, the anomaly may be increased by secondary density variations. In the case of limestone pinnacle reefs, which are petroleum-bearing structures in many parts of the world, compaction of the overlying formations can cause the density of these rocks to be slightly greater above the reef than to either side. A positive gravity anomaly is produced, which is the integrated attraction of anomalous mass over a considerable vertical interval, and this may be much greater than the effect of the reef itself.

Exploration for Minerals

The density of many ores of metallic minerals is considerably greater than that of normal crustal rock. Table 8.1 gives a few examples of important ones, with their densities.

TABLE 8.1

Mineral	Formula	Density (g/cm^3)
Pyrite	FeS_2	5·0
Pyrrhotite	$FeS(S_x)$	4·6
Chalcopyrite	$CuFeS_2$	4·2
Magnetite	Fe_3O_4	5·0
Galena	PbS	7·5
Chromite	$FeCr_2O_4$	4·5

Masses of these minerals, if they occur at reasonable depth, will produce detectable anomalies. Most of them can be located by other geophysical methods, and a gravity survey is not necessarily the most efficient means of prospecting for them. This is particularly true in areas of rugged terrain, where the calculation of the terrain correction is difficult, and even the determination of station elevation may be expensive.

The great advantage of gravity measurements is that the anomaly is a direct measure of mass, and it is total mass which is of commercial importance. It is shown in Appendix 1 that the surface integral of the normal attraction of any mass distribution, taken over any surface surrounding the mass, is equal to $4\pi GM$, where M is the total mass (theorem of Gauss). In the case of gravity measurements over an ore body, we do not have measurements over a closed surface, but only over a limited area of the earth's surface. However, provided the measurements extend to beyond the detectable limits of the local anomaly, the integral can be evaluated. Let the ore body be surrounded by the surface consisting of an infinite plane and hemisphere, as shown in Fig. 8.4. If the radius of the hemisphere is very large compared to the

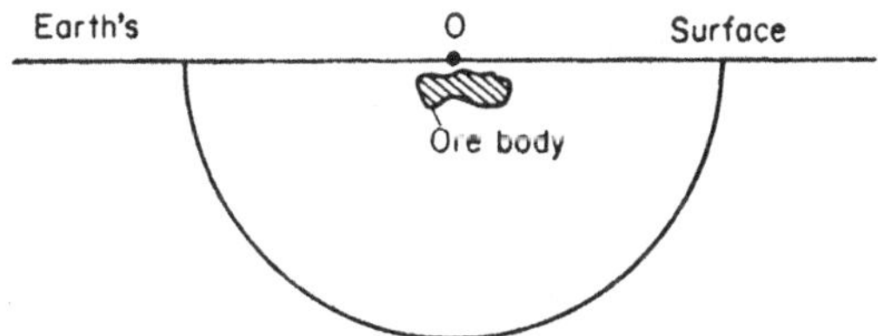

FIG. 8.4.   Surface chosen for Gaussian integration of the gravity field of an ore body.

dimensions of the body, the surface integral over it, of the inward attraction, reduces to $2\pi GM$. The integral over the plane, which is considered to coincide with the earth's surface, then must equal $2\pi GM$ also. In other words

$$\int \Delta g(x,y)\mathrm{d}x\mathrm{d}y = 2\pi GM. \tag{8.2}$$

Provided that the local anomaly can be isolated from the regional background, the integral on the left of equation (8.2) can be evaluated numerically from the residual anomaly contour map. The total anomalous mass M is thus determined uniquely, without any assumptions as to the form, position or density of the body. This fact is worth stressing, in view of the ambiguity of gravity interpretation as far as the *distribution* of the mass is concerned. The calculation of the actual ore mass, however, requires that the ore density ϱ and the density of surrounding rock ϱ_n be known:

$$\text{Actual mass} = M \times \frac{\varrho}{\varrho - \varrho_n}. \qquad (8.3)$$

In Fig. 8.5 is shown a profile across a pyrite body in northern Quebec, Canada (Goetz, 1958), with a form which gives the calculated profile plotted. This represents a rather ideal case, for the body occurred at shallow depth, and had the considerable vertical extent of 600 ft. The maximum residual anomaly of

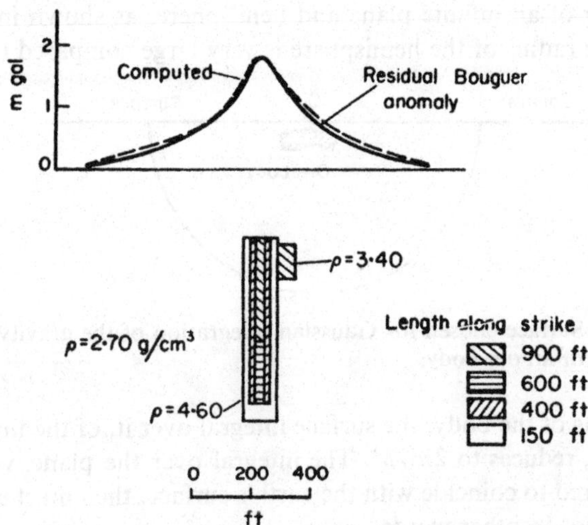

FIG. 8.5. Observed and computed anomalies, and deduced section, of a metallic ore body in Quebec (after Goetz).

$1 \cdot 7$ mgal is rather greater than could be expected in mineral exploration. Goetz, by integration of the residual anomaly according to equation (8.2), estimated that the total mass was about $3 \cdot 2 \times 10^6$ tons.

CHAPTER 9

Tidal Variation of Gravity

WE HAVE mentioned, in Chapter 2, that the value of g at any place on earth varies with time, because of the changing attractions of the sun and moon. As far as gravity surveys are concerned, this variation is usually considered together with instrumental drift as an extraneous effect which must be eliminated from the measurements. However, there is considerable interest in studying the tidal variation of gravity itself. If the earth were completely rigid, the variation would be completely determined by the positions of the sun and moon, and could be directly calculated. In fact, the earth yields slightly under the influence of the tidal forces, and the variation in g is slightly greater than that calculated for a rigid earth. The determination of this magnification of the tidal effect is thus of value in the study of the rigidity of the earth.

Tidal Accelerations

The theory of the accelerations produced by the sun and moon has been worked out in detail (Bartels, 1957), originally in connection with ocean tides. Let us consider first the lunar effect,

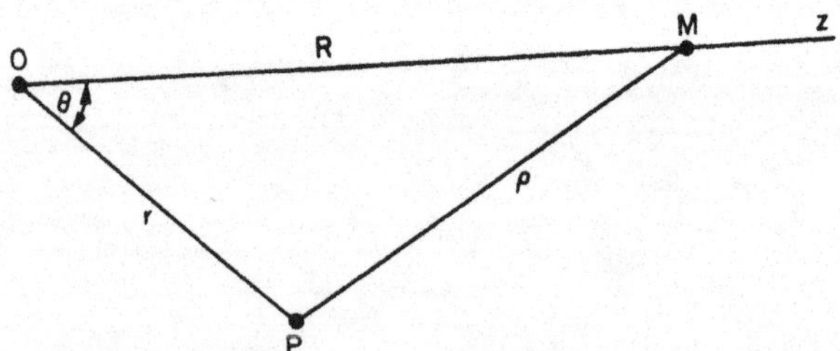

FIG. 9.1.   Relation between earth (centre O) and moon (centre M) for tidal calculations.

140

and begin by analysing the revolution of the earth and moon about their common centre of mass, but neglecting the earth's revolution on its axis. The centripetal acceleration of the earth and moon, required for the revolution, is provided by their mutual attraction, but only at the mass centre of each body is the gravitational force precisely equal to the centripetal force. In Fig. 9.1, O and M are the centres of mass of earth and moon respectively. At an arbitrary point P of the earth, the centripetal acceleration is the same as at O, and is also in the z-direction. The difference between this acceleration and the acceleration due to the moon's attraction, gives the tidal acceleration at P. The centripetal acceleration is derivable from a potential

$$\frac{GM}{R^2} \cdot z + C = \frac{GMr\cos\theta}{R^2} + C,$$

where C is a constant and M is the mass of the moon. At the point P, therefore, the moon's tidal potential is

$$U_m = GM\left(\frac{1}{\varrho} - \frac{r\cos\theta}{R^2} + C'\right)$$

or

$$U_m = GM\left(\frac{1}{\varrho} - \frac{1}{R} - \frac{r\cos\theta}{R^2}\right), \qquad (9.1)$$

where the constant is chosen to make the potential vanish at O.

The term $1/r$ can be expanded in a series of Legendre polynomials (Appendix 1), which gives

$$U_m = \frac{GMr^2}{R^3}\left[P_2(\cos\theta) + \frac{r}{R}P_3(\cos\theta) + \dots\right]. \qquad (9.2)$$

Since the ratio r/R is of the order of $1/60$, the first term in equation (9.2) is very often sufficient. It is usual to rewrite the equation in terms of the mean distance of the moon, c, as follows.

$$U_m = G(r)\left[\left(\frac{c}{R}\right)^3(\cos2\theta+\tfrac{1}{3})+\tfrac{1}{6}\frac{r}{c}\left(\frac{c}{R}\right)^4(5\cos3\theta+3\cos\theta)\right], \quad (9.3)$$

where

$$G(r) = \tfrac{3}{4}\,GM\frac{r^2}{c^3}.$$

Similarly, the tidal potential of the sun at a point on earth is

$$U_s = G_s(r)\left[\left(\frac{c_s}{R_s}\right)^3(\cos2\theta+\tfrac{1}{3})+\tfrac{1}{6}\frac{r}{c_s}\left(\frac{c_s}{R_s}\right)^4(5\cos3\theta+3\cos\theta)\right], \quad (9.4)$$

where

$$G_s(r) = \tfrac{3}{4}\,GS\frac{r^2}{c_s^3},$$

S is the mass of the sun, and distances with subscript s refer to the sun. In this case, r/c_s is of order $1/23{,}600$, and the first term in equation (9.4) gives a very close approximation to the total effect.

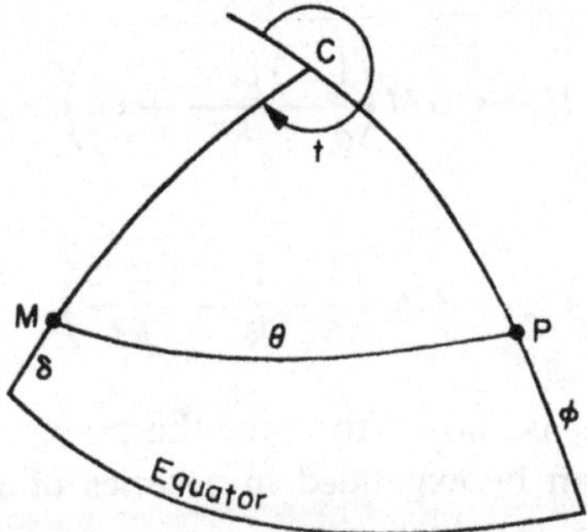

FIG. 9.2.   Definition of quantities involved in tidal calculations for a rotating earth.

If we now permit the earth to rotate, an observer at P moves through a varying potential field, and the potentials become functions of the time. In Fig. 9.2, C is the pole of the celestial sphere, M the moon, and P the point on earth. The moon's geocentral zenith distance θ, which appears in equation (9.3), can be expressed in terms of declination δ, geocentric latitude of P, φ, and t, the moon's hour angle:

$$\cos\theta = \sin\varphi\sin\delta + \cos\varphi\cos\delta\cos(t - 180°). \qquad (9.5)$$

Substitution of (9.5) into (9.3) gives, for the leading term,

$$U_m = G(r)\,(\tfrac{c}{R})^3[3(\tfrac{1}{3} - \sin^2\delta)(\tfrac{1}{3} - \sin^2\varphi) - \sin2\varphi\sin2\delta\cos t$$
$$+ \cos^2\varphi\cos^2\delta\cos2t]. \qquad (9.6)$$

An equivalent expression for U_s follows from (9.4), with δ_s, the sun's declination, and t_s, the sun's hour angle, in place of δ and t.

Components of acceleration at P are obtained by taking the appropriate derivatives of U_m and U_s, the variation in g being given by $-\partial U/\partial r$.

The three terms in equation (9.6) vary differently with time, and with the latitude of P. The first term varies only with δ and R, and leads to tides of long period. In the second term, the factor $\cos t$ indicates a period of one (lunar) day, corresponding to the diurnal tide, while the third term, containing $\cos 2t$, leads to the semidiurnal tide. Long period and semidiurnal tides are symmetric about the equator, but the diurnal tides are antisymmetric. Similarly, tides of three different periods arise from the sun's potential, with sidereal time appearing in place of lunar time.

Variation of Gravity on a Rigid Earth

The theory outlined above leads to a number of tidal components, of both diurnal and semidiurnal class. Table 9.1 lists the most important components, with their usual symbols and periods.

The dependence of the amplitude of the variation in g due to the different tides upon latitude is shown in Fig. 9.3. These curves show the amplitudes obtained directly from the expressions for the potential, and take no account of any yielding of the earth. They are therefore the tidal changes of g which would be observed on a completely rigid earth.

TABLE 9.1

Class	Symbol	Name	Period (hr)
Semidiurnal	M_2	Principal lunar	12·42
	S_2	Principal solar	12·00
	N_2	Lunar ellipticity	12·66
	K_2	Lunisolar	11·97
Diurnal	K_1	Lunisolar	23·93
	O_1	Lunar declination	25·82
	P_1	Solar declination	24·07

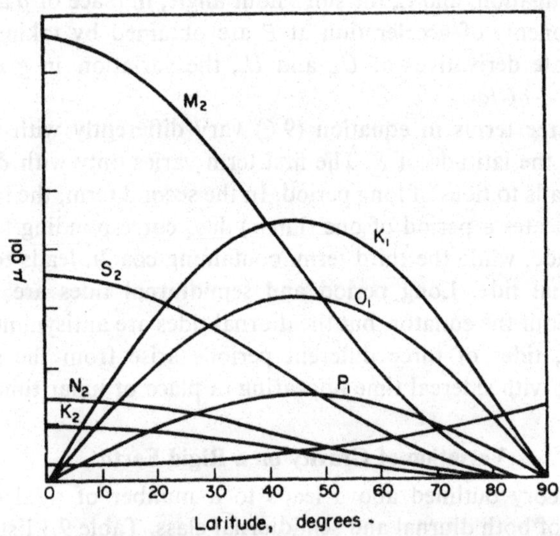

FIG. 9.3. Amplitudes of the tidal components as functions of latitude.

Variation of Gravity on a Yielding Earth

In treating tides in the solid earth, it is sufficient to assume that the displacements will follow an equilibrium theory. That is, for a tidal potential U, the surface elevation will be U/g. In this case, Love (1911) showed that for a tidal potential of second degree, the surface displacement and the potential due to the deformation are both proportional to it.

Consider a point on a yielding earth, where the radial displacement due to the earth tide is u. Then the total disturbed potential can be written

$$U = U_0 + U_t + U_m - gu, \tag{9.7}$$

where U_0 is the undisturbed potential due to attraction and rotation, U_t is the tidal potential, given by equations (9.3) and (9.4), U_m is the potential at the point due to the mass in the tidal deformation, and (gu) is the change in potential due to the displacement u.

Love introduced constants h and k such that

$$u = h(U_t/g),$$
$$U_m = kU_t. \tag{9.8}$$

In terms of h and k,

$$U = U_0 + U_t(1 + k - h). \tag{9.9}$$

The tidal contribution to g is obtained by taking the derivative of the variational terms in (9.7):

$$g_t = -\frac{\partial U_t}{\partial r} - \frac{\partial U_m}{\partial r} + \frac{u \partial g}{\partial r}. \tag{9.10}$$

Since U_t depends on r^{-2} (equation (9.3) or (9.4)),

$$\frac{\partial U_t}{\partial r} = -\frac{2U_t}{r}. \tag{9.11}$$

The term $\partial U_m/\partial r$ is the attraction of a mass coating which is distributed over the earth according to a second-degree harmonic. By equation (47) of Appendix 1,

$$\frac{\partial U_m}{\partial r} = -\frac{3}{r} U_m. \tag{9.12}$$

Finally, $\partial g/\partial r = -(2g/r)$ by the free-air formula. Substituting these values into equation (9.10) gives

$$g_t = [-2 - 2h + 3k]\,\frac{U_t}{r} \tag{9.13}$$

or, by (9.11),

$$g_t = [1 + h - \tfrac{3}{2}k]\,\frac{\partial U_t}{\partial r}. \tag{9.14}$$

If the earth were perfectly rigid, h and k would be identically zero, and we would have

$$g_t = \frac{\partial U_t}{\partial r}. \tag{9.15}$$

The factor $[1 + h - \tfrac{3}{2}k]$ therefore corresponds to the magnification of the tidal variations in gravity on a yielding earth. Love's numbers h and k also appear in the expression for the deflection of a horizontal pendulum due to tidal forces on a yielding earth. The combination of gravity and horizontal pendulum measurements allows h and k to be determined individually.

Earth–Tide Measurements

The importance of the above theory lies in the fact that h and k can be computed for various earth models. The gravimetric magnification factor $[1 + h - \tfrac{3}{2}k]$ for these models can therefore be compared with the factor obtained from observations, to indicate whether the elastic properties of the models are reasonable.

The usual approach is to make continuous observations of the variation in g at a fixed station, for a period of some weeks. Since the tidal variations are measured in microgals, gravimeters of high sensitivity, capable of producing continuous records, are required. Care is also necessary in the choice of the site, so that microseismic and other disturbances are reduced to a minimum. The observed variation in g (Fig. 9.4) is the composite effect of the main tidal components listed in Table 9.1, and the magnification factor must be obtained by comparison of the amplitude of each

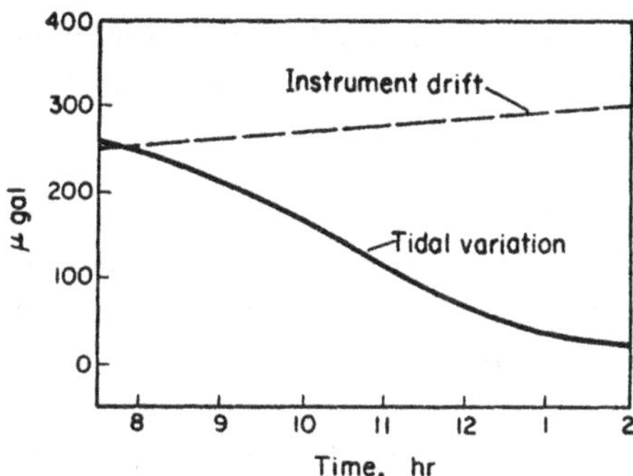

FIG. 9.4.   Portion of a tidal record made with a Lacoste gravimeter,
showing resolution to microgals.

component with the corresponding value for a rigid earth.
Harmonic or spectral analysis of the observed records is therefore
necessary, to separate the various components. Harmonic analysis
separates the diurnal and semidiurnal tides, but special techniques
of power spectral analysis, using numerical filters of very limited
width, are required to separate the components completely. The
analysis also yields the difference in phase (if any) between the
observed components and the theoretical tides.

In most of the earlier work the chief significance was assigned to
the semidiurnal components. This is partly because of the pre-
dominance of the M_2 tide, and also because the diurnal components have often been found to be contaminated by the effects of
heating and cooling of the crust in the vicinity of the observing
station. Apart from this, the chief difficulty in the interpretation of
the measurements has been the correction for a secondary effect
due to oceanic tides. The latter disturb the gravitational field
through the direct attraction of the water, and also through the
further distortion of the earth by the load of water. Methods have
been given (Egedal, 1959) for the calculation of the effect on
gravity, which can be used if oceanic tidal tables are available. The
largest effect is contributed by oceanic areas within a very few
degrees of a station.

The first attempt to determine the magnification factor from a series of simultaneous measurements made at stations distributed over the earth was carried out in 1949 (Baars, 1953). From two weeks of recordings at twenty-one stations, Baars obtained a value of $1 \cdot 24$ for the factor, with negligible phase difference from the theoretical tides. During the International Geophysical Year, many longer series of observations were made, and the sensitivity of gravimeters, such as the Askania and Lacoste–Romberg, was increased for the purpose. The more recent determinations (Pariisky, 1961; Nishimura, Ichinohe and Nakagawa, 1962) have given a mean value for the factor between $1 \cdot 138$ and $1 \cdot 142$, also with very small phase departure from the theoretical tides. There is some evidence for a difference in the factor determined from different tidal components, which is, in fact, predicted by calculations for model earths.

Comparison with Earth Models

A calculation of the values of Love's numbers to be expected on an earth yielding according to the equilibrium tide theory was given by Takeuchi (1950). For his calculations he assumed a number of earth models, with elastic properties in the core and mantle, adopted from the findings of seismology. The greatest difference between the different models was in the rigidity assumed for the core. Because the seismological evidence for the liquid state of the core is based on the absence of observed transverse waves through it, an independent test is very desirable. The calculations of Takeuchi showed that any modulus of rigidity between 0 and 10^9 dynes/cm^2 for the core would lead to values of h and k compatible with the observations, but larger values would not. For the earth model with zero rigidity in the core, Takeuchi obtained a value of $1 \cdot 184$ for the gravimetric factor. This is quite close to the mean of modern observations.

The equilibrium tide approximation is satisfactory in the case of the solid earth because the period of free oscillations is much less than that of the diurnal or semidiurnal tides. However, the effect of free oscillations cannot be neglected in the core, and a complete solution shows that Love's numbers are dependent upon the period of the tidal components. Solutions have been given by Jeffreys and Vicente (1957), Pekeris, Jarosch and Alterman (1959) and Molodensky (1961). The general result is that the gravimetric factor reaches a minimum for tidal periods very close to 24 hr (the K_1 component), and increases for tides of longer and shorter period. The difference between components such as M_2 or O_1 and K_1 varies between $0\cdot02$ and $0\cdot04$, depending on the earth model for which the calculations are based. Modern earth tide observations, subjected to careful spectral analysis, should detect these differences.

Other applications of earth tide measurements by means of gravimeters include the investigation of the anomalous tidal response of certain areas because of crustal fractures or other tectonic features (Tomaschek, 1957), and the measurement of gravity changes during eclipses. From time to time, suggestions have been made that gravitational fields can be absorbed by intervening mass, as electrostatic fields can be shielded by a conductor. If this were the case, the tidal effect of the sun or moon should be altered during an eclipse by the other body. All observations which have been made to date yield a negative result as far as any gravitational absorption is concerned.

For the correction of measurements made during gravity surveys, tables of the tidal variation are published at intervals in the journal *Geophysical Prospecting*. These are obtained by combining the theoretical rigid-earth tides with an assumed magnification factor, $1\cdot200$. Tables of this type can be used to reduce the number of repeat measurements required during a survey, provided the drift characteristics of the instrument itself are well known.

APPENDIX 1

The Elements of Potential Theory

The Potential Function

There is a broad class of force fields in which no dissipative losses of energy occur when a body is moved from one point to another. Fields of this type are said to be conservative, and the differences in potential energy of a body between points in them depend only on the positions of the points, not on the path along which the body moved.

Consider a conservative field of force in which the components of force on a unit mass at the point (x, y, z) are X, Y and Z, in the directions of Cartesian coordinates. Then if $_AW_B$ is the difference in potential energy of a mass m between two points A and B,

$$_AW_B = -m \int_A^B Xdx + Ydy + Zdz. \tag{1}$$

We define the difference in *potential* $_AU_B$, between A and B as the negative of the difference in potential energy per unit mass. Thus

$$_AU_B = \int_A^B Xdx + Ydy + Zdz. \tag{2}$$

It follows from (2) that at every point of the field

$$\left. \begin{aligned} X &= \frac{\partial U}{\partial x} \\[2mm] Y &= \frac{\partial U}{\partial y} \\[2mm] Z &= \frac{\partial U}{\partial z} \end{aligned} \right\}, \tag{3}$$

or, in vector notation, the vector field $\mathbf{F}$ is given by

$$\mathbf{F} = \text{grad } U. \tag{4}$$

Equation (2) defines differences in potential, but not the absolute value. We choose a function for U which vanishes at infinity. The great advantage of formulating problems in terms of the potential, rather than the vector field, is that U is a scalar, and the knowledge of only one value at each point is required to define it.

Of fundamental importance in the gravitational case is the potential of a mass particle. By Newton's law of gravitation, the field at distance r from a mass m is

$$\mathbf{F} = \frac{Gm}{r^2}. \tag{5}$$

Applying (2), and integrating from infinity to the point at distance r, we have

$$U = \frac{Gm}{r}. \tag{6}$$

It should be pointed out that there is not universal agreement on the choice of signs in potential theory. The above convention is the one usually adopted in geodesy and geophysics, as it is convenient to treat the earth's potential as a positive quantity. In other applications, the potential is given the same sign as the potential energy.

Potentials of Spherical Distributions

Spherical shell (Fig. 1). If the surface density of mass is σ, the potential at P is

$$U = G\sigma \int \frac{ds}{\varrho}, \tag{7}$$

where ds is an element of surface.

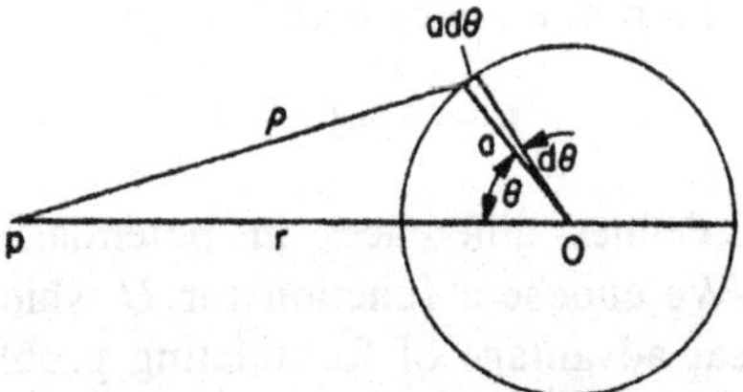

FIG. 1.   Attraction of a spherical shell.

If the figure is rotated about OP, a ring is traced out on the shell. Integration over the range of the variable θ adds the contributions of all such rings, so that

$$U = 2\pi a^2 \sigma G \int_0^\pi \frac{\sin\theta\,d\theta}{\varrho}. \tag{8}$$

Since

$$\varrho^2 = a^2 + r^2 - 2ar\cos\theta, \qquad \frac{\sin\theta\,d\theta}{\varrho} = \frac{d\varrho}{ra}$$

and

$$U = \frac{2\pi a\sigma G}{r} \int_{\varrho_1}^{\varrho_2} d\varrho, \tag{9}$$

where ϱ_1 and ϱ_2 are the least and greatest values of ϱ.

If P is outside the shell,

$$\varrho_2 - \varrho_1 = 2a$$

and

$$U = \frac{4\pi a^2 \sigma G}{r}$$

$$= \frac{Gm}{r}, \tag{10}$$

where m is the mass of the shell.

The potential is identical to that of a particle of mass m at O.

Since a solid sphere can be considered as built up of an infinite number of shells, its external potential is also identical to that of a central particle.

If P is inside the shell,

$$\varrho_2 - \varrho_1 = 2r$$

and

$$U = 4\pi a \sigma G$$

$$= \frac{MG}{a}, \tag{11}$$

which is constant.

The field is therefore zero everywhere within a spherical shell.

General Properties of Potential Fields

A number of important relationships follow from the fact that

$$\mathbf{F} = \text{grad } U.$$

We note first that

$$\text{curl } \mathbf{F} = 0 \text{ everywhere} \tag{12}$$

for the components of curl $\mathbf{F}$ contain terms of the form

$$\left[\frac{\partial^2 U}{\partial y \partial z} - \frac{\partial^2 U}{\partial z \partial y}\right],$$

which are identically zero, since U is a continuous function of the space coordinates.

The concept of lines of force of the field is useful. We imagine lines to be constructed such that at any point the field vector is tangent to a line of force, and the intensity of the force in any direction is measured by the number of lines crossing a unit area taken perpendicular to that direction. The total number of lines of force crossing any surface is known as the flux of the field.

The creation of flux within a volume V, which is surrounded by a

F

surface S, may easily be expressed as a volume integral throughout V:

$$\int_V \left(\frac{\partial X}{\partial x} + \frac{\partial Y}{\partial y} + \frac{\partial Z}{\partial z}\right)\mathrm{d}v = \int_V \mathrm{div}\mathbf{F}\mathrm{d}v \tag{13}$$

or as a surface integral over S,

$$\int_S F_n \mathrm{d}s,$$

where F_n is the component of F along the outward normal to S. If S encloses no attracting matter,

$$\int_V \mathrm{div}\mathbf{F}\mathrm{d}v = \int_S F_n \mathrm{d}s = 0. \tag{14}$$

Equation (14) remains valid if S shrinks about any point of V, and this is possible only if

$$\mathrm{div}\ \mathbf{F} = 0 \quad \text{everywhere.}$$

Since $\mathbf{F} = \mathrm{grad}\ U$,

$$\mathrm{div}\ \mathrm{grad}\ U = 0$$

or

$$\nabla^2 U = \frac{\partial^2 U}{\partial x^2} + \frac{\partial^2 U}{\partial y^2} + \frac{\partial^2 U}{\partial z^2} = 0 \tag{15}$$

at all points of V.

Equation (15) is known as Laplace's equation, and is the fundamental relationship of potential theory.

If a spherical surface S encloses an attracting particle of mass m at the centre, we may easily evaluate

$$\int_S F_n \mathrm{d}s.$$

It is

$$\int_S F_n \mathrm{d}s = -4\pi Gm. \tag{16}$$

Equation (16) is obviously valid for any arbitrary surface which lies outside the sphere, because there would be no change in flux between the two surfaces. It is thus possible to superimpose the effects of all mass within any surface S, and to write

$$\int_S F_n ds = -4\pi GM. \tag{17}$$

where M is the total enclosed mass. Equation (17) is the theorem of Gauss.

In terms of the volume integral,

$$\int_V \operatorname{div} F dv = -4\pi GM. \tag{18}$$

If the volume V is allowed to shrink about a point of V, we have

$$\operatorname{div} \mathbf{F} = -4\pi G\varrho, \tag{19}$$

where ϱ is the density at the point.

Thus,

$$\nabla^2 U = -4\pi G\varrho, \tag{20}$$

which is Poisson's equation.

Development of the Potential in Series

For many computational problems, a series expansion of the potential is essential. Consider (Fig. 2) the expression for the potential at P due to a unit mass at Q. The potential U is given by

$$U = G \cdot \frac{1}{p}. \tag{21}$$

Normally, however, the positions of P and Q are given with respect to the origin O.

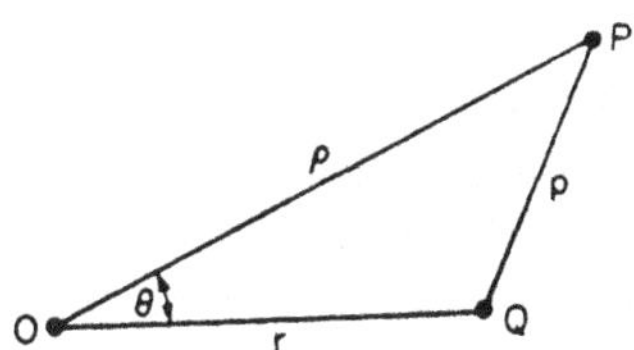

FIG. 2.   Quantities involved in series expansion of potential.

Since

$$p^2 = r^2 + \varrho^2 - 2r\varrho\cos\theta \tag{22}$$

we may write

$$\frac{1}{p} = \frac{1}{\varrho}\left(1 - 2\frac{r}{\varrho}\cos\theta + \frac{r^2}{\varrho^2}\right)^{-1/2}$$

$$= \frac{1}{\varrho}(1 - 2u\cos\theta + u^2)^{-1/2}, \tag{23}$$

where $u = r/\varrho$.

Provided $|2u\cos\theta - u^2| < 1$, we may expand the right side of equation (23), to give

$$\frac{1}{p} = \frac{1}{\varrho}\Big[P_0(\cos\theta) + P_1(\cos\theta)u + P_2(\cos\theta)u^2 + \ldots\Big], \tag{24}$$

where

$$P_0(\cos\theta) = 1$$

$$P_1(\cos\theta) = \cos\theta$$

$$P_2(\cos\theta) = \tfrac{3}{2}(\cos^2\theta - \tfrac{1}{3})$$

and in general

$$P_n(\cos\theta) = \sum_{k=0}^{n/2} \frac{1.3\ldots(2n-2k-1)}{2^k k!(n-2k)!}(-1)^k \cos^{n-2k}\theta. \tag{25}$$

The functions $P_n(\cos\theta)$ are known as Legendre polynomials.

Boundary Value Problems in Potential Theory

In a great number of problems of geodesy and geophysics, we must find a solution of Laplace's equation when values of the potential, or the field, are given over some surface. Recognition of the following theorem is essential for an appreciation of these problems.

We consider a volume V, surrounded by a surface S, and assume that Laplace's equation

$$\nabla^2 U = 0 \tag{15}$$

is satisfied throughout V. Let us assume that there are two solutions, U_1 and U_2, of (15), which take on the same values for all points of S. We may then show that U_1 and U_2 are identical at all points of V.

We take $U = U_1 - U_2$, and note that U also satisfies (15), and is zero on S.

An easily established identity is

$$\left(\frac{\partial U}{\partial x}\right)^2 + \left(\frac{\partial U}{\partial y}\right)^2 + \left(\frac{\partial U}{\partial z}\right)^2 = \operatorname{div}(U \operatorname{grad} U) - U \nabla^2 U. \tag{26}$$

Writing this identity for each point of V, and integrating, gives

$$\int_V \left[\left(\frac{\partial U}{\partial x}\right)^2 + \left(\frac{\partial U}{\partial y}\right)^2 + \left(\frac{\partial U}{\partial z}\right)^2 \right] dv$$

$$= \int_V \operatorname{div}(U \operatorname{grad} U) dv - \int_V U \nabla^2 U dv. \tag{27}$$

The second integral on the right side of equation (27) vanishes, because $\nabla^2 U$ is everywhere zero. The first may be transformed by Green's theorem on surface and volume integrals to

$$\int_S U \operatorname{grad} U ds.$$

But since U vanishes on S, this is also zero. Thus, at every point of V

$$\left(\frac{\partial U}{\partial x}\right)^2 + \left(\frac{\partial U}{\partial y}\right)^2 + \left(\frac{\partial U}{\partial z}\right)^2 = 0. \tag{28}$$

The difference $U_1 - U_2$ must therefore be constant, but since it is zero on S it must be zero everywhere. The theorem is thus established that solutions of Laplace's equation satisfying the same boundary conditions are identical.

If, on S, $\partial U_1/\partial n = \partial U_2/\partial n$, instead of $U_1 = U_2$, the above argument may be easily modified to show that U_1 and U_2 can differ only by a constant throughout V, and the force fields corresponding to U_1 and U_2 are again identical.

Plane Sheet of Mass Equivalent to a Given Field

An important application of the above theorem is the determination of a distribution of mass over a surface which produces a given field at all points outside of the surface. The simplest case occurs when the surface is an infinite horizontal plane between the observer and the attracting mass.

It is shown in Chapter 5 that the normal attraction of a plane sheet of mass of constant surface density σ is

$$\Delta g = G\sigma\Omega, \tag{29}$$

where Ω is the solid angle subtended by the sheet. If σ is not constant, but is a continuous function of coordinates (x, y) in the plane, the attraction at point on the sheet is

$$\Delta g(x,y) = 2\pi G\sigma(x,y). \tag{30}$$

This result is easily obtained by considering a small circle about the point $P(x,y)$, of such a radius that σ does not differ from σ_p by more than ε. As the observer approaches the sheet, the solid angle subtended by the circle approaches 2π, regardless of the radius of the circle. It is apparent that, in the limit, equation (30) will be satisfied.

A surface distribution, of density $\Delta g(x,y)/2\pi G$, spread over any horizontal plane on which Δg is known, will therefore produce the same field at all higher points as the actual masses which are responsible for the field Δg. This result is the basis for the surface integral formulae used in the calculation of derivatives of the field in Chapter 5.

Laplace's Equation in Spherical Coordinates

The spheroidal shape of the earth lends itself to the use of

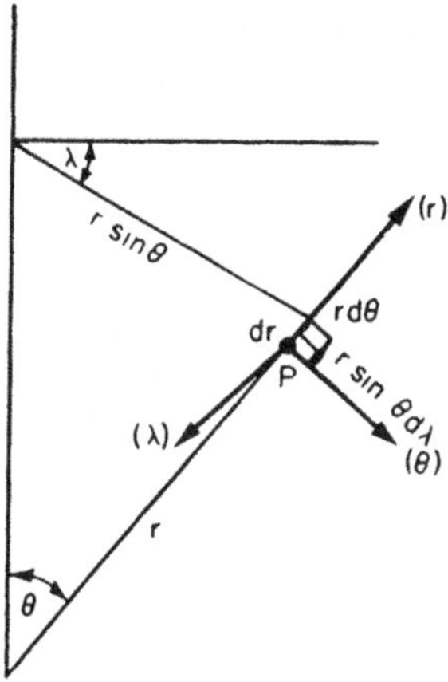

FIG. 3.   A volume element in spherical polar coordinates.

spherical polar coordinates, rather than Cartesian coordinates, for many applications. We take coordinates r, θ and λ as in Fig. 3. The components of force, in the three perpendicular directions shown, are the space gradients of U:

$$\frac{\partial U}{\partial r}, \qquad \frac{1}{r}\frac{\partial U}{\partial \theta}, \qquad \text{and} \qquad \frac{1}{r\sin\theta}\frac{\partial U}{\partial \lambda}.$$

An elementary cube about the point P has edges dr, $rd\theta$ and $r\sin\theta d\lambda$. If we write the divergence theorem, equation (14), for this cube, we obtain

$$\sin\theta\,\frac{\partial}{\partial r}\left(r^2\frac{\partial U}{\partial r}\right) + \frac{\partial}{\partial \theta}\left(\sin\theta\,\frac{\partial U}{\partial \theta}\right) + \frac{1}{\sin\theta}\frac{\partial^2 V}{\partial \lambda^2} = 0, \qquad (31)$$

which is Laplace's equation in spherical coordinates.

Writing $u = \cos\theta$, we have

$$\left(1-u^2\right)^{1/2}\frac{\partial}{\partial r}\left(r^2\frac{\partial U}{\partial r}\right) + \frac{\partial}{\partial u}\left[(1-u^2)\frac{\partial U}{\partial u}\right]$$

$$+ \frac{1}{(1-u^2)^{1/2}}\frac{\partial U}{\partial \lambda} = 0. \qquad (32)$$

If we try a solution in the form

$$U = r^n p(\theta) q(\lambda), \tag{33}$$

when n in an integer, and p and q are functions respectively of θ and λ only, we find

$$n(n+1) + \frac{1}{p(\theta)} \frac{\mathrm{d}}{\mathrm{d}u}\left[(1-u^2)\frac{\mathrm{d}p(\theta)}{\mathrm{d}n}\right] + \frac{1}{1-u^2}\frac{1}{q(\lambda)}\frac{\mathrm{d}^2 q(\lambda)}{\mathrm{d}\lambda^2} = 0. \tag{34}$$

The fact that only the third term contains λ shows that

$$\frac{1}{q(\lambda)}\frac{\mathrm{d}^2 q(\lambda)}{\mathrm{d}\lambda^2}$$

is constant. Therefore, $q(\lambda)$ must be of the form

$$q(\lambda) = A\cos m\lambda + B\sin m\lambda, \tag{35}$$

where m is an integer, and A and B are constants. With this substitution into (34), we find that $p(\theta)$ satisfies

$$\frac{\mathrm{d}}{\mathrm{d}n}\left[(1-u^2)\frac{\mathrm{d}p(\theta)}{\mathrm{d}n}\right] + \left[n(n+1) - \frac{m^2}{1-u^2}\right]p(\theta) = 0, \tag{36}$$

which is Legendre's associated equation.

Solutions of equation (36) are functions $P_n^m(u)$, where

$$P_n^m(u) = \frac{(n-m)!}{2^n(n!)^2}\sin^m\theta\left(\frac{\mathrm{d}}{\mathrm{d}u}\right)^{n+m}(u^2-1)^n. \tag{37}$$

The solution of Laplace's equation is then

$$U = r^n P_n^m(u)(A\cos m\lambda + B\sin m\lambda), \tag{38}$$

which is a solid spherical harmonic of degree n.

It is not difficult to show that a second solution is

$$U = r^{-(n+1)} P_n^m(u)(A\cos m\lambda + B\sin m\lambda). \tag{39}$$

The choice of the solutions (38) or (39) in a particular problem depends on whether the potential is to remain finite at the origin or at infinity.

If there is no dependence upon longitude, $m = 0$, and the solution is given by

$$U = r^n P_n(u) \tag{40}$$

or

$$U = r^{-(n+1)} P_n(u).$$

Here, $P_n(u)$ is the Legendre polynomial introduced in equation (25), and is known as a zonal harmonic. The more general function $(A\cos m\lambda + B\sin m\lambda)P_n^m(u)$ (for $m < n$) is known as a tesseral harmonic; it has the property of vanishing on m meridians of longitude, and $(n-m)$ parallels of latitude.

The functions $P_n^m(u)$ for low values of m and n are:

$$\left.\begin{aligned}
P_1(\cos\theta) &= \cos\theta \\
P_1^1(\cos\theta) &= \sin\theta \\
P_2(\cos\theta) &= \tfrac{1}{2}(3\cos^2\theta - 1) \\
P_2^1(\cos\theta) &= 3\sin\theta\cos\theta \\
P_2^2(\cos\theta) &= 3\sin^2\theta
\end{aligned}\right\} . \tag{41}$$

Potential of a Coating on a Sphere

We consider a surface density $\sigma(\theta,\lambda)$ on a sphere of radius a, with centre at the origin O, and we assume that $\sigma(\theta,\lambda)$ can be expanded as a series of surface spherical harmonics:

$$\sigma(\theta,\lambda) = \sum_{n=0}^{\infty} S_n(\theta,\lambda), \tag{42}$$

where $S_n(\theta,\lambda)$ is of the form $(A\cos m\lambda + B\sin m\lambda)P_n^m(\cos\theta)$. The potential at an external point $P(r, \theta, \lambda)$ is

$$U = \int_0^{2\pi} \int_{-1}^{1} \frac{G\sigma a^2 du_1 d\lambda_1}{\varrho}, \tag{43}$$

where the subscript 1 refers to a particular element of surface, at Q, $u = \cos\theta$, $\varrho = PQ$, and γ is the angle between OP and OQ.

We have
$$\varrho^2 = r^2 + a^2 - 2ar\cos\gamma$$

and
$$\cos\gamma = \cos\theta\cos\theta_1 + \sin\theta\sin\theta_1\cos(\lambda - \lambda_1). \tag{44}$$

For an external point, $1/\varrho$ can be expanded as in equation (24). Interchanging the summation and integration gives

$$U = \sum_{n=0}^{\infty} G\frac{a^{n+2}}{r^{n+1}} \int_0^{2\pi} \int_{-1}^{1} \sigma P_n(\cos\gamma)\mathrm{d}n_1\mathrm{d}\lambda_1. \tag{45}$$

Comparison of equation (45) with equation (39) shows that the integral

$$\int_0^{2\pi} \int_{-1}^{1} \sigma P_n(\cos\gamma)\mathrm{d}u_1\mathrm{d}\lambda_1$$

must be a surface spherical harmonic of degree n.

In fact, substituting the expansion for σ, and the value for $\cos\gamma$ from (44) permits evaluation of the integral to yield (MacRobert, 1947, p. 137)

$$\int_0^{2\pi} \int_{-1}^{1} \sigma P_n(\cos\gamma)\mathrm{d}u_1\mathrm{d}\lambda_1 = \frac{4}{2n+1} S_n(\theta,\lambda). \tag{46}$$

So that

$$U = 4\pi G \sum_{n=0}^{\infty} \frac{1}{2n+1} \frac{a^{n+2}}{r^{n+1}} S_n(\theta,\lambda). \tag{47}$$

A similar argument shows that the potential inside the sphere is

$$U_i = 4\pi G \sum_{n=0}^{\infty} \frac{1}{2n+1} \frac{r^n}{a^{n-1}} S_n(\theta,\lambda). \tag{48}$$

If the potential is given on a sphere of radius a as a series of surface spherical harmonics, the above development may be reversed to give an integral expression for the potential at any external point:

$$U = \frac{a(r^2 - a^2)}{4\pi} \int\limits_0^{2\pi} \int\limits_{-1}^{1} \frac{U(\theta_1\lambda_1)}{\varrho^{3/2}} \, \mathrm{d}u_1 \mathrm{d}\lambda_1 \tag{49}$$

or at an internal point:

$$U_i = \frac{a(a^2 - r^2)}{4\pi} \int\limits_0^{2\pi} \int\limits_{-1}^{1} \frac{U(\theta_1\lambda_1)}{\varrho^{3/2}} \, \mathrm{d}u_1 \mathrm{d}\lambda_1. \tag{50}$$

APPENDIX 2

Equilibrium Form of a Rotating Fluid

FOR a rotating mass of fluid to be in equilibrium under the forces of its own attraction and those of rotation, the external surface, and the surfaces of equal pressure within the mass, must be equipotentials. If the density of the fluid is constant, an oblate ellipsoid of revolution is a possible form for the external surface, but only for a certain range of values of the angular velocity of rotation. A great deal of attention was given during the nineteenth century to this problem (Darwin, 1910).

We consider a body of fluid rotating about the z-axis with angular velocity ω. Over the external surface, we must have

$$U + \tfrac{1}{2}\omega^2(x^2 + y^2) = \text{constant}, \tag{1}$$

where U is the potential produced by the static mass. The problem is to determine if this surface can be the ellipsoid

$$\frac{x^2}{a^2} + \frac{y^2}{b^2} + \frac{z^2}{c^2} = 1 \tag{2}$$

for the particular case $a = b$. The calculation of U involves the transformation of Laplace's equation to ellipsoidal coordinates; it is a classical problem and will not be repeated here. The internal potential (Kellogg, 1929, p. 194) for a tri-axial ellipsoid is

$$U_i = \pi G \varrho a^2 c \, [C_0 - C_1 x^2 - C_2 y^2 - C_3 z^2], \tag{3}$$

where

$$C_0 = \int_0^\infty \varphi(s)^{-1/2} \mathrm{d}s$$

$$C_1 = \int_0^\infty (a^2 + s)^{-1} \, \varphi(s)^{-1/2} \mathrm{d}s$$

164

$$C_2 = \int_0^\infty (b^2 + s)^{-1}\, \varphi(s)^{-1/2} ds$$

$$C_3 = \int_0^\infty (c^2 + s)^{-1}\, \varphi(s)^{-1/2} ds$$

and
$$\varphi(s) = (a^2 + s)(b^2 + s)(c^2 + s).$$

The expression for U can be put in terms of normal elliptic integrals by the transformations

$$c = a\cos\gamma, \quad \varkappa = \sqrt{\frac{a^2 - b^2}{a^2 - c^2}}, \quad (\varkappa_1 = 1 - \varkappa^2). \qquad (4)$$

Then
$$U = \pi\varrho Gabc\left[\frac{2}{a\sin\gamma}F + \frac{2}{a^3\sin^3\gamma}\left\{\frac{x^2}{\varkappa^2}(E - F) + y^2\left(\frac{\sin\gamma\cos\gamma}{\varkappa_1^2}\right.\right.\right.$$

$$\left.\left.\left. + \frac{F}{\varkappa^2} - \frac{E}{\varkappa^2\varkappa_1^2}\right) + \frac{z^2}{\varkappa_1^2}(E - \tan\gamma)\right\}\right], \qquad (5)$$

where

$$E = \int_0^\gamma (1 - \varkappa^2\sin^2\gamma)^{1/2} d\gamma$$

and

$$F = \int_0^\gamma (1 - \varkappa^2\sin^2\gamma)^{-1/2} d\gamma.$$

The potential U is now in the form

$$U = Ax^2 + By^2 + Cz^2 + D, \qquad (6)$$

and for this to be consistent with (1) and (2),

$$a^2(A + \tfrac{1}{2}\omega^2) = b^2(B + \tfrac{1}{2}\omega^2) = a^2 C. \qquad (7)$$

For the ellipsoid of revolution, $A = B$, and $\varkappa = 0$. Then

$$\tfrac{1}{2}\omega^2 = C\cos^2\gamma - A \qquad (8)$$

and the evaluation of the integrals leads to

$$\omega^2 = \frac{2\pi G\varrho}{\tan^3\gamma}\left[\gamma(3+\tan^2\gamma)-3\tan\gamma\right]. \tag{9}$$

There is no real solution of (9) if

$$\frac{\omega^2}{2\pi G\varrho} > 0\cdot 2247.$$

For a body with density ϱ equal to the mean density of the earth, the limiting value of ω gives a period of rotation of $2\frac{1}{2}$ hr. For rotation with shorter period, an ellipsoid of revolution is not a possible form for the fluid mass.

When $\omega^2/2\pi G\varrho < 0\cdot 2247$, equation (9) yields two values of $\tan\gamma$ for a given ω, and the corresponding flattening f of the two possible ellipsoids is obtained from

$$f = 1-\cos\gamma. \tag{10}$$

One solution normally has a large value of flattening, and the other a much smaller value. For a body with uniform density $5\cdot 5$ g/cm³, and ω equal to the earth's angular velocity, the latter solution has a flattening of about 1/230.

When the theory is extended to include a variation of density within the fluid, the departure from ellipsoidal shape, mentioned in Chapter 3, is indicated, and the flattening is found to be much less for given mean density and ω.

The satisfactory calculation of the theoretical or hydrostatic flattening of the earth requires a knowledge of the actual external potential field with a precision that has only become available through a study of satellite orbits. Using satellite data, Jeffreys (1963) estimated that, if the earth were in hydrostatic equilibrium, the flattening would be 1/299·67. Caputo (1965), by holding different parameters fixed, obtained a range of values for the hydrostatic flattening, but he suggested that 1/299·49 was most reliable. These values are less than the value of the actual flattening (1/297 implied by the International Gravity Formula, 1/282·2 indicated by satellite observations), a fact which shows that the earth as a whole has considerable long-term strength.

The flattening of surfaces of equal density (which are equipotentials) within the fluid decreases toward the centre. For example, Bullard (1948) obtained a value of 1/390 for the flattening of the boundary of the earth's core, on the hydrostatic theory.

Determination of the External Field from Satellite Observations

THE analysis of the dynamics of satellite orbits follows well-established principles of celestial mechanics (Sterne, 1960), with complications imposed by the effect of air resistance (Kaula, 1962).

To describe the motion of the satellite in space, geocentric coordinates x, y and z are used (Fig. 1), with the x and y in the equatorial plane. The angle Ω of Fig. 1, which defines the intersection of the plane of orbit with the x–y plane, is known as the

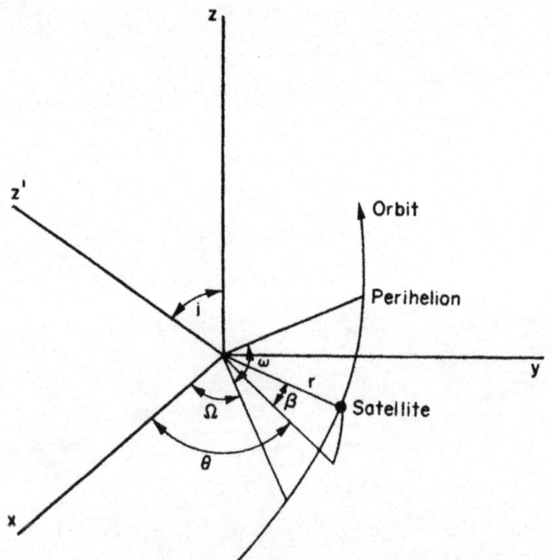

FIG. 1.   Quantities defining a satellite orbit.

longitude of ascending node, while i is the inclination of the orbit. The form and orientation of the elliptical orbit in its plane are specified by the eccentricity e, and the argument of perihelion ω. Finally, the size of the orbit is usually given in terms of mean distance a.

The position of the satellite in orbit can be given in terms of polar coordinates (r, β, θ) where β is geocentric declination, and θ is right ascension. If the satellite describes the orbit with period T, its mean motion n is defined to be $2\pi/T$.

It is usual to set up the differential equations in terms of the Hamiltonian $H(p_r, p_\beta, p_\theta, q_r, q_\beta, q_\theta, L)$ where L is the kinetic potential, the q's are coordinates, and the p's are momenta defined by

$$p_j = \frac{\partial L}{\partial \dot{q}}. \tag{1}$$

The details can be found in any work on advanced mechanics. In particular, the Hamiltonian H is

$$H = \sum p_j \dot{q}_j - L \tag{2}$$

over all coordinates, and the systems of equations of motion are

$$\frac{\partial H}{\partial p_j} = \dot{q}_j, \quad \frac{\partial H}{\partial \dot{q}_j} = -\dot{p}_j. \tag{3}$$

If the earth's external field contained only a potential of the form

$$U = \frac{GM}{r} \tag{4}$$

the Hamiltonian would be found to be

$$H_0 = \tfrac{1}{2}\left(p_r^2 + \frac{p_\theta^2}{r^2\cos^2\beta} + \frac{p_\beta^2}{r^2}\right) - \frac{GM}{r} \tag{5}$$

and the resulting differential equations would lead to unperturbed elliptical motion, obeying Kepler's laws.

As we have seen, the first-order departure of the field from spherical symmetry involves the second harmonic, and U is of the form

$$U = \frac{GM}{r}\left[1 - \frac{B}{r^2}(3\cos^2\beta - 1) + \ldots\right]. \qquad (6)$$

The second term in equation (6) represents the perturbing Hamiltonian in the first-order theory. Solution of the differential equations with this perturbation yields the important result that the elements Ω and ω are subject to secular variations:

$$\left.\begin{aligned}
\dot{\Omega} &= -\frac{3Bn_0}{a_0^2(1-e_0^2)}\cos i_0 \\[2ex]
\dot{\omega} &= \frac{3Bn_0}{a_0^2(1-e_0^2)^2}(2 - \tfrac{5}{2}\sin^2 i_0)
\end{aligned}\right\}, \qquad (7)$$

where the subscript 0 refers to initial values. Thus, both Ω and ω vary, unless $(2 - \tfrac{5}{2}\sin^2 i_0)$ vanishes. The direct connection between these secular variations, which, amounting to several degrees per day, are easily observed, and the constant B, led to the relatively early success of satellite observations in revising the estimate of the second harmonic term in U, and therefore of the earth's flattening.

Higher harmonics in the earth's field lead to periodic variations in the elements, which can be detected by precise tracking. Kaula (1962) has described the procedure for these cases.

Bibliography

AIRY, G. B. (1855) *Phil. Trans. Roy. Soc. Lond.* **145**, 101.

BAARS, B. (1953) *Geophys. Prosp.* **1**, 82.

BARTELS, JULIUS (1957) *Handbuch der Physik* **48**, 734.

BEALS, C. S., FERGUSON, G. M., and LANDAU, A. (1956) *J. Roy. Astr. Soc. Can.* **50**, 203.

BOTT, M. H. P. (1961) *Geophys. J.* **5**, 207.

BOUGUER, P. (1749) *La figure de la terre*, Paris.

BOWIE, W. (1917) *U.S. Coast and Geod. Surv.*, Spec. Pub. 40.

BROWNE, B. C. (1937) *Mon. Not. Roy. Astr. Soc.*, Geophys. Suppl. **4**, 271.

BULLARD, E. C. (1936) *Phil. Trans. Roy. Soc. Lond.* A, **235**, 445.

BULLARD, E. C. (1948) *Mon. Not. Roy. Astr. Soc.*, Geophys. Suppl. **5**, 186.

BULLEN, K. E. (1953) *An Introduction to the Theory of Seismology*, Camb. Univ. Press, Cambridge.

CAPUTO, MICHELE (1965) *J. Geophys. Res.* **70**, 955.

CLAIRAUT, A. C. (1743) *Théorie de la figure de la terre*, Paris.

CLARK, J. S. (1940) *Phil. Trans. Roy. Soc. Lond.* A, **238**, 65.

COOK, A. H. (1952) *Mon. Not. Roy. Astr. Soc.* Geophys. Suppl. **6**, 243.

COOK, A. H. and MURPHY, T. (1952) Dublin Inst. for Advanced Studies, *Geophys. Mem.*, No. 2, Pt. 4.

COOK, A. H. (1957) *Preparations for a New Absolute Determination of Gravity at the National Physical Laboratory, Teddington.* Commun. from the Nat. Phys. Lab., Teddington.

COOK, KENNETH L. (1962) *Advances in Geophysics*, **9**, Chap. 6, Academic Press, New York.

DARWIN, SIR G. H. (1910) *Scientific Papers*, Camb. Univ. Press, Cambridge.

DEAN, WILLIAM C. (1958) *Geophysics* **23**, 97.

DUTTON, C. E. (1889) *Bull. Phil. Soc. Wash.* **11**, 51.

EGEDAL, J. (1959) *J. Atmos. Terr. Phys.* **16**, 318.

ELKINS, T. A. (1951) *Geophysics* **16**, 29.

EVJEN, H. M. (1936) *Geophysics* **1**, 127.

EWING, J., and EWING, M. (1959) *Bull. Geol. Soc. Am.* **70**, 291.

GARLAND, G. D., KANASEWICH, E. R., and THOMPSON, THOMAS L. (1961) *J. Geophys. Res.* **66**, 2495.

GAY, MALCOLM W. (1940) *Geophysics* **5**, 176.

GOETZ, JOSEPH F. (1958) *Geophysics* **23**, 606.

GRAF, ANTON, and SCHULZE, REINHARD (1961) *J. Geophys. Res.* **66**, 1813.

GRANT, FRASER (1952) *Geophysics* **17**, 344 and 756.

GULATEE, B. L. (1940) *Survey of India*, Professional Paper 30.

GUNN, ROSS (1937) *J. Franklin Inst.* **224**, 19.

GUNN, ROSS (1945) *Trans. Amer. Geophys. Union of 1944*, Part IV, 633.

HALES, A. L. (1935) *Mon. Not. Roy. Astr. Soc.*, Geophys. Suppl. **3**, 372.

HAMMER, SIGMUND (1939) *Geophysics* **4**, 184.

HARRISON, J. C. (1956) *Phil. Trans. Roy. Soc. Lond.* A, **248**, 283.

HASKELL, N. A. (1935) *Physics* **6**, 265.

HAYFORD, J. F., and BOWIE, WILLIAM (1912) *U.S. Coast and Geod. Surv.*, Spec. Pub. 10.

HEEZEN, B. C. (1960) *Scientific American* **203**, 98.

HEISKANEN, W. A. (1924) *Publ. Finnish Geod. Inst.*, No. 4.

HEISKANEN, W. A. (1938) *Publ. Isostat. Inst., Int. Ass. Geod.*, No. 1, Helsinki.

HEISKANEN, W. A. (1957) *Trans. Amer. Geophys. Un.* **38**, 841.

HEISKANEN, W. A., and VENING MEINESZ, F. A. (1958) *The Earth and its Gravity Field.*, McGraw-Hill, New York.

HELMERT, F. A. (1884) *Die mathematischen und physikalischen Theorie der höheren Geodäsie*, Teubner, Leipzig.

HERSCHEL, SIR JOHN (1849) *Outlines of Astronomy*, London.

HESS, HARRY H. (1938) *Proc. Amer. Phil. Soc.* **79**, 71.

HEYL, PAUL R. (1930) *J. Res. Nat. Bur. Stand.* **5**, 1243.

HEYL, PAUL R., and COOK, GUY S. (1936) *J. Res. Nat. Bur. Stand* **17**, 805.

HONKASALO, TAUNO (1959) *Geofisica* **7**, 2.

INNES, M. J. S. (1957) *Trans. Amer. Geophys. Un.* **38**, 156.

INNES, M. J. S. (1960) *Publ. Domin. Obs.*, Ottawa, **21**, No. 6.

INNES, M. J. S. (1961) *J. Geophys. Res.* **66**, 2225.

JACKSON, J. E. (1961) *Geophys. J.* **4**, 375.

JEFFREYS, H. (1943) *Mon. Not. Roy. Astr. Soc.*, Geophys. Suppl. **5**, 55.

JEFFREYS, H. (1948) *Mon. Not. Roy. Astr. Soc.*, Geophys. Suppl. **5**, 219.

JEFFREYS, H. (1952) *The Earth* (3rd Ed.), Camb. Univ. Press, Cambridge.

JEFFREYS, H., and VICENTE, R. O. (1957) *Mon. Not. Roy. Astr. Soc.* **117**, 157, 165.

JEFFREYS, H. (1963) *Geophys. J.* **8**, 196.

JUNG, K. (1952) Gerland's Beit. zur *Geophysik* **62**, 39.

JUNG, K. (1952b) *Landolt-Börnstein Zahlenwerte und Funktionen* **3**, 265, Springer, Berlin.

KATER, H. (1818) *Phil. Trans. Roy. Soc. Lond.* **108**, 32.

KAULA, W. M. (1959) U.S. Army Map Service Tech. Report 24, Washington.

KAULA, W. M. (1961) *J. Geophys. Res.* **66,** 1799.

KAULA, W. M. (1962) *Advances in Geophysics,* **9,** Chap. 5, Academic Press, New York.

KELLOGG, OLIVER DIMON (1929) *Foundations of Potential Theory,* Ungar, New York.

KÜHNEN, F., and FURTWÄNGLER, P. (1906) *Veröff. Press. Geodät. Inst.* **27.**

LACOSTE, L. J. B. (1934) *Physics* **5,** 178.

LACOSTE, LUCIEN (1959) *Geophysics* **24,** 309.

LACOSTE, L. J. B., and HARRISON, J. C. (1961) *Geophys. J.* **5,** 89.

LAMBERT, W. D., and DARLING, F. W. (1931) *Bull. géod.,* No. 32.

LAMBERT, W. D., and DARLING, F. W. (1936) *U.S. Coast and Geod. Surv.,* Spec. Pub., 199.

LOMNITZ, C. (1959) *Proc. Fifth World Petrol Congress,* New York.

LOVE, A. E. H. (1911) *Some Problems of Geodynamics,* Camb. Univ. Press, Cambridge.

MACE, C. (1939) *Mon. Not. Roy. Astr. Soc., Geophys. Suppl.,* **4,** 473.

MACROBERT, T. M. (1947) *Spherical Harmonics* (2nd Ed.), Methuen, London.

MASKELYNE, NEVIL (1774) *Phil. Roy. Soc. Lond.* **50,** 495.

MERSON, R. H., and KING-HELE, D. G. (1958) *Nature* **182,** 640.

MOLODENSKY, M. S. (1961) Quatrième Symposium Int. sur les marées terrestres. (*Comm. de l'Observatoire Royale de Belgique,* No. 188, 25.)

MOLODENSKY, M. S., EREMEEV, V. F., and YURKINA, M. I. (1960) *Methods for Study of the External Gravitational Field and Figure of the Earth,* Moscow. (Translation published by Israel Program for Scientific Translations, Jerusalem, 1962.)

MUNK, W. H., and MACDONALD, G. J. F. (1960) *The Rotation of the Earth,* Camb. Univ. Press, Cambridge.

NAFE, J. E., and DRAKE, C. L. (1957) Society of Exploration Geophysicists, Annual Meeting. Unpublished paper. The graph referred to is published in Steinhart and Meyer (1961) and Talwani, Sutton and Worzel (1959).

NETTLETON, L. L. (1939) *Geophysics* **4,** 176.

NETTLETON, L. L. (1940) *Geophysical Prospecting for Oil,* McGraw-Hill, New York.

NETTLETON, L. L. (1942) *Geophysics* **7,** 293.

NETTLETON, L. L., LACOSTE, LUCIEN, and HARRISON, J. C. (1960) *Geophysics* **25,** 181.

NISHIMURA, E., ICHINOHE, T., and NAKAGAWA, I. (1962) *Report on Earth Tidal Observations of Gravity with the Askania Gravimeter No. 111 in Japan during the International Geophysical Year 1957–58.* Science Council of Japan, Ueno Part, Tokyo.

NISKANEN, E. (1939) *Publ. Isostat. Inst.*, Helsinki, No. 6.

O'KEEFE, JOHN A., ECKELS, ANN, and SQUIRE, R. KENNETH (1959) *Astr. J.* **64**, 245.

PAKISER, L. C., PRESS, FRANK, and KANE, M. R. (1960) *Bull. Geol. Soc. Amer.* **71**, 415.

PARIISKY, N. N. (1961) Quatrième Symposium Int. sur les marées terrestres. (*Commun. de l'Observatoire Royale de Belgique*, No. 188, 96.)

PEKERIS, C. L. (1935) *Mon. Not. Roy. Astr. Soc.*, Geophys. Suppl. **3**, 343.

PEKERIS, C. L., JAROSCH, H., and ALTERMAN, Z. (1959) *Third Int. Symposium on Earth Tides* (Pub. of the Istituto di Topografia e Geodesia dell' Universita, Trieste), 17.

PETERS, LEO J. (1949) *Geophysics* **14**, 290.

PIZETTI, P. (1911) *Atti della Reale Acad. della Scienze di Torino* **46**.

PRATT, J. H. (1855) *Phil. Trans. Roy. Soc. Lond.* **145**, 53.

PRATT, J. H. (1859) *Phil. Trans. Roy. Soc. Lond.* **149**, 745.

PRESTON-THOMAS, H., TURNBULL, L. G., GREEN, E., DAUPHINEE, T. M., and KALRA, S. N. (1960) *Can. J. Phys.* **38**, 824.

RAYLEIGH, 3RD BARON (1916) *Phil. Mag.* **32**, 529.

RICE, DONALD A. (1951) *Deflections of the vertical from gravity anomalies.* U.S. Coast and Geod. Surv., Washington.

RUSSELL, R. D., JACOBS, J. A., and GRANT, F. S. (1960) *Bull. Geol. Soc. Amer.* **71**, 1223.

SOLLINS, A. D. (1947) *Bull. Géod.*, N.S., **6**, 279.

STEINHART, JOHN S., and MEYER, ROBERT P. (1961) *Carnegie Inst. Wash. Publ.* 622.

STERNE, THEODORE E. (1960) *An Introduction to Celestial Mechanics*, Interscience Publ., New York.

STOKES, G. G. (1849) *Trans. Camb. Phil. Soc.* **8**, 672.

SUTTON, G. H. (1960) Program of Int. Assoc. Seism. and Phys. of Earth's Interior, Helsinki General Assembly (abstract).

TAKEUCHI, H. (1950) *Trans. Amer. Geophys. Un.* **31**, 651.

TALWANI, MANIK, and EWING, MAURICE (1960) *Geophysics* **25**, 203.

TALWANI, MANIK, HEEZEN, BRUCE C., and WORZEL, J. LAMAR (1961) *Publ. Cen. Seism. Bur. Int. Un. Geod. and Geophys., Travaux Scient.* **22**, 81.

TALWANI, MANIK, SUTTON, GEORGE H., and WORZEL, J. LAMAR (1959) *J. Geophys. Res.* **64**, 1545.

TALWANI, MANIK, WORZEL, J. LAMAR, and EWING, MAURICE (1961) *J. Geophys. Res.* **66**, 1265.

TOMASCHEK, RUDOLF (1957) *Handbuch der Physik* **48**, 775.

TSUBOI, CHUJI (1938) *Proc. Imp. Acad. Tokyo* **14**, 170.

UOTILA, U. A. (1957) Pub. 7, Ohio State Univ., Inst. of Geodesy, Photogrammetry and Cartography, Columbus.

Vening Meinesz, F. A. (1929) *Theory and Practice of Gravity Measurements at Sea*, Delft.

Vening Meinesz, F. A. (1931) *Bull. géod.* No. 29.

Vening Meinesz, F. A. (1941) *Publ. Netherlands Geod. Comm.*, Delft.

Vening Meinesz, F. A. (1954) *Bull. Geol. Soc. Amer.* **65**, 143.

Volet, Ch. (1952) *C.R. Acad. Sci. Paris* **235**, 442.

Woollard, G. P. (1950) *Geophysics* **15**, 1.

Woollard, G. P. (1959) *J. Geophys. Res.* **64**, 1521.

Worzel, J. L., and Shurbet, G. L. (1955) *Geol. Soc. Amer. Spec. Pap.* 62, 87.

Yungul, S. H. (1961) *Geophysics* **26**, 45.

Zhongolovich, I. D. (1952) *Publ. Inst. Theor. Astr., Acad. Sci. U.S.S.R.*, Moscow.

Zhongolovich, I. D. (1956) *Publ. Inst. Theor. Astr.* 6, *Acad. Sci. U.S.S.R.*, Moscow.

Subject Index

Made in the USA
Monee, IL
07 July 2026

56545361R00108